**CHEMICAL
BACKGROUND
FOR THE
BIOLOGICAL
SCIENCES**

PRENTICE-HALL FOUNDATIONS
OF MODERN BIOLOGY SERIES

WILLIAM D. MCELROY
AND CARL P. SWANSON, *editors*

CHEMICAL BACKGROUND FOR THE BIOLOGICAL SCIENCES

2nd edition

EMIL H. WHITE
Professor of Chemistry, The Johns Hopkins University

PRENTICE-HALL, INC.

ENGLEWOOD CLIFFS, NEW JERSEY

FOUNDATIONS OF MODERN BIOLOGY SERIES   WILLIAM D. MCELROY
AND CARL P. SWANSON, *editors*

C—13-128447-9
P—13-128439-8
Library of Congress Catalog Card Number   77–91831

Current printing   10 9 8 7 6 5 4 3 2 1

PRENTICE-HALL INTERNATIONAL, INC., *London*
PRENTICE-HALL OF AUSTRALIA, PTY. LTD., *Sydney*
PRENTICE-HALL OF CANADA, LTD., *Toronto*
PRENTICE-HALL OF INDIA PRIVATE LTD., *New Delhi*
PRENTICE-HALL OF JAPAN, INC., *Tokyo*

FOREWORD

THIS SERIES, FOUNDATIONS OF MODERN BIOLOGY, WHEN LAUNCHED A NUMBER
of years ago, represented a significant departure in the organization of instruc-
tional materials in biology. The success of the series provides ample support for
the belief, shared by its authors, editors, and publisher, that student needs for
up-to-date, properly illustrated texts and teacher prerogatives in structuring a
course can best be served by a group of small volumes so planned as to en-
compass those areas of study central to an understanding of the content, state,
and direction of modern biology. The twelve volumes of the series still represent, in our view, a meaningful division of subject matter.

   This edition thus continues to reflect the rapidly changing face of biology;
and many of the consequent alterations have been suggested by the student
and teacher users of the texts. To all who have shown interest and aided us
we express thankful appreciation.

WILLIAM D. MCELROY
CARL P. SWANSON

THIS VOLUME IS A BRIEF OUTLINE OF CHEMISTRY. IT WAS WRITTEN FOR THE student of biochemistry and biology, and for others wishing an introduction to the subject. Included are concise, systematic treatments of atomic structure and inorganic chemistry, and a somewhat lengthier treatment of organic chemistry; applications of chemistry to biology have been given wherever possible. A short summary of the metric system and of the use of exponents and logarithms completes the volume. The approach throughout has been to emphasize the most important aspects of chemistry and to treat selected topics in depth, consistent with the limited space available. As in any introductory work, the use of a large number of new terms has been necessary. These are indicated in italics at the point of first use and they are listed in the index to facilitate the location of the definitions. Drawings, figures, and equations have been freely used to illustrate points in the text; they have been designed as integral parts of the volume and they should be studied as carefully as the material in the text. The author hopes that this attempt to show how and why chemical substances react will enable the reader to get a fuller and richer understanding of biochemistry and biology; the aim has been to show that chemistry is, in fact, one of the foundations of biology.

The organization of the first edition along chemical lines has been retained in the second edition as the most logical under the circumstances. The present volume should allow considerable flexibility in usage: some readers will use only the chemical material, either at the introductory level or as a review, and some will choose a bare minimum of the chemical subjects as a background for biological courses. Still others will be interested in the borderline area between chemistry and biology.

To increase the value of this book to the latter group of users, a number of biological-chemical relationships have been included. Some biological phenomena have been added for their intrinsic interest, some to show that underlying molecular processes are often responsible for biological facts, and still others to show that biological processes are often illuminated by the examination of simpler but related chemical processes. Principally for the latter reason, greater emphasis is placed on stereochemistry and reaction mechanisms than is usual in texts aimed at biologists. The application of chemical concepts can only enrich biology since the underlying physical processes are the same for the two sciences.

Because of the small size of this book, it has been necessary to omit many worthwhile subjects. In particular, the reader is urged to read other volumes in this series for more detailed information on the structure and properties of proteins, and on the nucleic acids and their role in genetics; the function of light in photosynthesis and in other biological processes is also covered elsewhere in the series.

EMIL H. WHITE

CONTENTS

CHEMICAL
BACKGROUND
FOR THE
BIOLOGICAL
SCIENCES

CHEMISTRY IS THE SCIENCE THAT DEALS WITH THE COMPOSITION AND STRUCTURE of matter and with the transformations that matter undergoes. Chemistry is a rather broad field; at one extreme, in theoretical chemistry and spectroscopy, it borders on physics, and at the other extreme, in organic chemistry, it borders on biochemistry and biology. The problems that chemists work on reflect this broad range; in recent years, for example, chemists have published research papers on a mathematical model of methane, the generation of chemical elements by the reaction of high-energy protons with iron, the infrared detection of water in the atmosphere of Venus, the preparation of compounds of xenon and krypton, the action of light on benzene, the mechanism of the oxidation of ethyl alcohol, the determination of the formula and structure of firefly luciferin, the synthesis of penicillin, and studies of the mode of action of the enzyme chymotrypsin. Despite the apparent variety of these investigations, however, chemistry is a body of knowledge united by a relatively small number of basic principles; these principles are the subject of the present volume.

Chemistry became a science during the latter part of the eighteenth century, largely as a result of the application of quantitative methods to chemical phenomena by the French scientist Antoine Lavoisier. It is proper, therefore, that we begin this volume with a section on the measurement of matter.

The *quantity* of matter in any given system is the *mass* of that system, where mass is measured in terms of a standard unit, the *kilogram* (kg). The *weight* of an object is a measure of the force of gravity on it. Thus, a man who weighs 160 lb on the earth would weigh 32 lb on the surface of the moon; his mass, of course, would be the same on both planets. Two other important standard units of measurement in the sciences are that of length, the *meter* (m), and that of time, the *second* (sec). These units have been variously defined in the past. At the present time, a kilogram is defined as a mass equal to that of a block of platinum-iridium alloy kept at Sèvres, France. The meter is defined as a length equal to 1,650,753.73 times the wavelength of the orange-red light emitted by the element krypton (Kr^{86}). The second, until recently, was defined as a time interval equal to 1/31,556,925.9747 of the tropical year beginning January 1900, but it has now been redefined as the time required for the nucleus of the isotope $_{55}Ce^{133}$ to change its orientation in a magnetic field 9,192,631,770 times.

Two important subunits of this kilogram-meter-second system of weights and measures are the *gram* (g), which is $\frac{1}{1000}$ of a kilogram, and the *centimeter* (cm), which is $\frac{1}{100}$ of a meter.

The kilogram, meter, and second are the fundamental units of the *metric system*° of measurement, a system used by scientists throughout the world. The metric system is also the standard system of weights and measures of many countries. In the commercial affairs of the United States and Great Britain, however, the British system of measurement is used; the fundamental units of this system are the pound, the foot, and the second.

A casual examination of the outward appearance of an object usually leads one to conclude that matter is continuous in nature. That is, the average person would conclude from such an observation that a block of a substance such as pure copper could, in principle, be subdivided an infinite number of times. This view of matter is incorrect. If the copper block were to be divided enough times, a stage would ultimately be reached at which no further division could be effected. Similarly, the division of other substances would yield other indivisible particles. These indivisible particles, the building blocks of matter, are called *atoms.* Of course, atoms can be split in nuclear reactions involving extremely large energy changes, but so long as we restrict ourselves to chemical reactions and simple physical changes, our notion of indivisible atoms is correct.

Atoms are exceedingly small. A block of ordinary copper weighing 63.5 g

° See Appendix A.

(slightly more than 2 ounces) contains 602,000,000,000,000,000,000,000 atoms of copper;° the reason for picking this particular weight of copper and number of atoms will be given in a later section. This large a number cannot be comprehended meaningfully. Suppose that we had 6.02×10^{23} peas; what volume would they occupy? Assuming that the peas were average in size (about 100, or 10^2 peas per cubic inch), how many peas would completely fill a household refrigerator? About 10^6. How many would fill an ordinary house from cellar to attic? 10^9. How many would be required to fill all the houses in a city the size of Chicago? 10^{15}. How many would be required to form a uniform layer 10 ft deep over the entire surface of the earth? 10^{22}. At this point, most of our peas still remain! To use up 6.02×10^{23} peas, we would have to blanket with 10 ft of peas about 60 planets the size of the earth.†

We are dealing here with very large numbers, and it is the very large number of atoms in even a microscopic piece of matter that gives the erroneous impression that matter is continuous. We see now that it would be impossible to divide a copper block down to its constituent atoms with a knife or some other crude device. Using more subtle approaches, however, scientists can readily detect and work with individual atoms in the cyclotron and in the mass spectrometer, and in fact, most of the information we have concerning atoms comes from such experiments.

A substance composed of a single kind of atom is called an *element;* copper, iron, and oxygen, for example, are elements. It is a surprising fact that all the matter in the earth is made up of just 103 elements. All of the compounds with which we are familiar derive from various combinations of these elements. Actually, a few of the elements are synthetic, or man-made, and others are quite rare. About 98 percent of the mass of the earth is composed of just seven elements; these are, in order of decreasing abundance, iron, oxygen, silicon, magnesium, nickel, calcium, and aluminum.

Atoms are not hard, homogeneous spheres. Experiments with high-energy particles have shown that atoms are complex systems made up of a number of smaller particles; of these, the *electron* (e), the *proton* (p), and the *neutron* (n) are the most important. The atoms of the 103 elements are composed of different numbers and proportions of these three fundamental particles. The proton, which bears a positive electrical charge, and the neutron, which bears no charge, have approximately the same mass. On the other hand, the electron, which has a negative charge equal in magnitude to the positive charge on the

---

° Expressed in powers of 10, this number becomes 6.02×10^{23}; see Appendix B.
† Adapted from D. H. Andrews and R. J. Kokes, *Fundamental Chemistry* (New York: John Wiley & Sons, Inc., 1962), pp. 9–10.

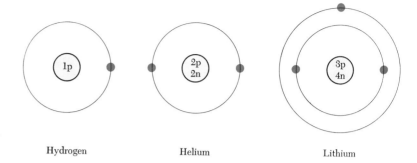

Figure 1.1 **Planetary models of various atoms.**

Hydrogen   Helium   Lithium

proton, has a mass of only 1/1,836 that of the proton. Since atoms are electrically neutral, they must contain equal numbers of electrons and protons.

Our understanding of the way these fundamental particles are arranged in the atom is due largely to the work of the English physicist Ernest Rutherford and the Danish physicist Niels Bohr. In the model of the atom developed by these scientists, the protons and neutrons are concentrated in a small volume at the center of the atom (called the *nucleus*), and the electrons move in regular, defined orbits about the nucleus. This model is usually referred to as the planetary model of the atom in view of its superficial resemblance to the solar system.

The smallest and simplest atom of all is the hydrogen atom, which consists of a single proton and a single electron. The structure of this atom is given in Figure 1.1 along with the structures of the second and third lightest atoms, helium and lithium. Note that the sizes of the nuclei are exaggerated in the figures; the nucleus of the hydrogen atom has a diameter of about 10^{-13} cm, whereas the diameter of the atom itself is about 100,000 times larger (10^{-8} cm). The helium atom consists of two protons, two neutrons, and two electrons, and the most common type of lithium atom consists of three protons, four neutrons, and three electrons. If this listing of the fundamental particles is continued for the remaining elements, a graded series is formed, in which the nuclei of the

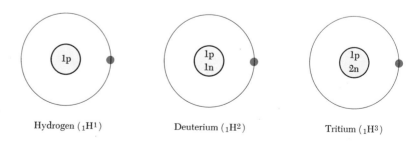

Figure 1.2 **The three isotopes of hydrogen.**

Hydrogen ($_1H^1$)   Deuterium ($_1H^2$)   Tritium ($_1H^3$)

atoms contain from 1 to 103 protons plus a slightly larger number of neutrons. The number of protons in the nucleus is called the *atomic number* of the element; for our three examples, then, the atomic numbers are 1, 2, and 3, respectively. A convenient way of representing the elements consists of the symbol for that element, a subscript giving the atomic number of the element, and a superscript giving the total number of protons and neutrons in the nucleus. Thus, atoms of the three elements in Figure 1.1 are designated by the symbols $_1H^1$, $_2He^4$, and $_3Li^7$. These symbols also refer, in a general way, to massive amounts of the elements themselves.

ISOTOPES   Although ordinary hydrogen is largely $_1H^1$, careful analyses have demonstrated the presence of two other varieties of the element with slightly different physical properties. The atomic structures for these three varieties are shown in Figure 1.2. It can be seen that they differ only in the number of neutrons in the nucleus. Forms of an element with the same nuclear charge but with different numbers of neutrons in the nucleus are called *isotopes;* other examples are $_3Li^6$, $_3Li^7$; $_6C^{12}$, $_6C^{14}$; and $_8O^{16}$, $_8O^{17}$, and $_8O^{18}$. All of the isotopes of an element undergo identical chemical reactions, although the rates of these reactions differ slightly. The isotopes of hydrogen each bear a specific name (deuterium and tritium), but those of the heavier elements do not.

ATOMIC WEIGHT   A list of the 103 elements, along with their symbols, atomic numbers, and atomic weights, is given in Table 1.1. The atomic weights are derived in the following way. It would be inconvenient to use the absolute weights of the atoms in laboratory work; the hydrogen atom weighs 1.67×10^{-24} g, for example, and it would be impossible to measure out such a small quantity of matter. Therefore, a set of useful weights has been devised for the elements in which relative atomic weights are assigned to the elements, starting with a value near 1 for hydrogen (the lightest element), in an order parallel to the absolute weights of their atoms. The scale has been based on several different elements in the past. At the present time, the *atomic weight* of an element is defined as the weight of an average atom of that element relative to the weight of a $_6C^{12}$ atom assigned the integral value $\underline{12}$.° This relationship is illustrated in Table 1.2. The units for the atomic weight scale are called atomic mass units (AMU); thus, the atomic weight of hydrogen is 1.00797 AMU, that of helium is 4.0026 AMU, and that of lithium is 6.939 AMU. The mass of the proton on the $_6C^{12}$ scale is 1.0073 AMU and the mass of the neutron is 1.0087 AMU. Since these masses are very close to the value 1 and since the mass of the electron is small enough to be neglected, the atomic weights of individual isotopes should have values that are very nearly integral. Many elements, as they occur in nature, consist of largely one isotope, and the atomic weights of these elements are in fact very close to integers (for example, H, He, N, O, and F; see Table 1.1). Other elements, as they occur in nature, consist of mixtures of iso-

° $\underline{12}$ = 12.0000 . . .; similarly, other integral values are indicated by an underscore.

Table 1.1 Table of atomic weights (based on carbon 12)[a]

ELEMENT	SYMBOL	ATOMIC NUMBER	ATOMIC WEIGHT	ELEMENT	SYMBOL	ATOMIC NUMBER	ATOMIC WEIGHT
Actinium	Ac	89	[227][d]	Mercury	Hg	80	200.59
Aluminum	Al	13	26.9815	Molybdenum	Mo	42	95.94
Americium	Am	95	[243][d]	Neodymium	Nd	60	144.24
Antimony	Sb	51	121.75	Neon	Ne	10	20.183
Argon	Ar	18	39.948	Neptunium	Np	93	[237][d]
Arsenic	As	33	74.9216	Nickel	Ni	28	58.71
Astatine	At	85	[210][d]	Niobium	Nb	41	92.906
Barium	Ba	56	137.34	Nitrogen	N	7	14.0067
Berkelium	Bk	97	[249][d]	Nobelium	No	102	[254][d]
Beryllium	Be	4	9.0122	Osmium	Os	76	190.2
Bismuth	Bi	83	208.980	Oxygen	O	8	15.9994[b]
Boron	B	5	10.811[b]	Palladium	Pd	46	106.4
Bromine	Br	35	79.909[c]	Phosphorus	P	15	30.9738
Cadmium	Cd	48	112.40	Platinum	Pt	78	195.09
Calcium	Ca	20	40.08	Plutonium	Pu	94	[242][d]
Californium	Cf	98	[251][d]	Polonium	Po	84	[210][d]
Carbon	C	6	12.01115[b]	Potassium	K	19	39.102
Cerium	Ce	58	140.12	Praseodymium	Pr	59	140.907
Cesium	Cs	55	132.905	Promethium	Pm	61	[147][d]
Chlorine	Cl	17	35.453[c]	Protactinium	Pa	91	[231][d]
Chromium	Cr	24	51.996[c]	Radium	Ra	88	[226][d]
Cobalt	Co	27	58.9332	Radon	Rn	86	[222][d]
Copper	Cu	29	63.54	Rhenium	Re	75	186.2
Curium	Cm	96	[247][d]	Rhodium	Rh	45	102.905
Dysprosium	Dy	66	162.50	Rubidium	Rb	37	85.47
Einsteinium	Es	99	[254][d]	Ruthenium	Ru	44	101.07
Erbium	Er	68	167.26	Samarium	Sm	62	150.35
Europium	Eu	63	151.96	Scandium	Sc	21	44.956
Fermium	Fm	100	[253][d]	Selenium	Se	34	78.96
Fluorine	F	9	18.9984	Silicon	Si	14	28.086[b]
Francium	Fr	87	[223][d]	Silver	Ag	47	107.870[c]
Gadolinium	Gd	64	157.25	Sodium	Na	11	22.9898
Gallium	Ga	31	69.72	Strontium	Sr	38	87.62
Germanium	Ge	32	72.59	Sulfur	S	16	32.064[b]
Gold	Au	79	196.967	Tantalum	Ta	73	180.948
Hafnium	Hf	72	178.49	Technetium	Tc	43	[99][d]
Helium	He	2	4.0026	Tellurium	Te	52	127.60
Holmium	Ho	67	164.930	Terbium	Tb	65	158.924
Hydrogen	H	1	1.00797[b]	Thallium	Tl	81	204.37
Indium	In	49	114.82	Thorium	Th	90	232.038
Iodine	I	53	126.9044	Thulium	Tm	69	168.934
Iridium	Ir	77	192.2	Tin	Sn	50	118.69
Iron	Fe	26	55.847[c]	Titanium	Ti	22	47.90
Krypton	Kr	36	83.80	Tungsten	W	74	183.85
Lanthanum	La	57	138.91	Uranium	U	92	238.03
Lawrencium	Lw	103	[257][d]	Vanadium	V	23	50.942
Lead	Pb	82	207.19	Xenon	Xe	54	131.30
Lithium	Li	3	6.939	Ytterbium	Yb	70	173.04
Lutetium	Lu	71	174.97	Yttrium	Y	39	88.905
Magnesium	Mg	12	24.312	Zinc	Zn	30	65.37
Manganese	Mn	25	54.9380	Zirconium	Zr	40	91.22
Mendelevium	Md	101	[256][d]				

[a] Adapted from *Chemical and Engineering News*, November 20, 1961, p. 43.

[b] The atomic weight varies because of natural variations in the isotopic composition of the element. The observed ranges are boron, ±0.003; carbon, ±0.00005; hydrogen, ±0.00001; oxygen, ±0.0001; silicon, ±0.001; sulfur ±0.003.

[c] The atomic weight is believed to have an experimental uncertainty of the following magnitude: bromine, ±0.002; chlorine, ±0.001; chromium, ±0.001; iron, ±0.003; silver, ±0.003. For other elements the last digit given is believed to be reliable to ±0.5.

[d] A value given in brackets is the mass number of the isotope of longest known lifetime.

Table 1.2 **Atomic weights of the light elements**[a]

ELEMENT	ABSOLUTE WEIGHT OF A REPRESENTATIVE ATOM	POSSIBLE SCALE (WEIGHT RELATIVE TO HYDROGEN = $\underline{1}$)	PRESENT SCALE (WEIGHT RELATIVE TO $_6C^{12} = \underline{12}$)[b]
Hydrogen	1.6737×10^{-24} g	$\underline{1}$	1.00797
Helium	6.646×10^{-24} g	3.9709	4.0026
Lithium	11.52×10^{-24} g	6.884	6.939
Beryllium	14.964×10^{-24} g	8.9406	9.0122
Boron	17.951×10^{-24} g	10.725	10.811
Carbon	19.945×10^{-24} g	11.916	12.0115
$_6C^{12}$ isotope	19.925×10^{-24} g	11.905	$\underline{12}$

[a] Including their isotopes.

[b] See Table 1.1 for a complete listing.

topes. Ordinary magnesium, for example, contains 78.7 percent of $_{12}Mg^{24}$, 10.1 percent of $_{12}Mg^{25}$, and 11.2 percent of $_{12}Mg^{26}$; the atomic weight of magnesium, 24.312, is a proportional average of the atomic weights of the isotopes present. The atomic weight of ordinary carbon is 12.01115, not $\underline{12}$, for the same reason.

Atomic weights are useful to the chemist in that they permit him to work in a quantitative fashion with weighable quantities of matter. A weight in grams of an element numerically equal to the atomic weight of that element is called a *gram atomic weight,* or a *gram atom.* A gram atom of each and every element contains the same number of atoms since the atomic weights are assigned in proportion to the masses of single atoms. The number of atoms in a gram atom of an element, 6.02252×10^{23}, is known as *Avogadro's number.*

In practice, the symbols we use for the elements have three meanings. The symbol Li, for example, stands for the element lithium in a general way; it stands for a single atom of Li; and it also stands for a gram atom of lithium. If we know that 1 atom of Li reacts with 1 atom of H_2, then we know from our definitions that 1 gram atom of Li will react with 1 gram atom of H_2, and also, from our values of the atomic weights (Table 1.1), that 6.939 g of Li will react with exactly 1.008 g of H_2.

In 1896, the French scientist Antoine Henri Becquerel found that the element uranium and its salts produced an image on a photographic plate and that this exposure could take place through thick layers of paper and other materials. Shortly afterward, Marie and Pierre Curie discovered two other elements capable of the same action, polonium and radium. Since then, many others have been discovered. We recognize now that elements of this type, called *radioactive elements*, emit one or more of three types of penetrating rays: alpha (α) rays, which are composed of streams of α particles identical to helium atoms minus their two planetary electrons (in other words, helium nuclei); beta (β) rays, composed of electrons; and gamma (γ) rays, composed of short-wavelength X rays. These rays originate in the nuclei of the radioactive atoms. The nucleus is thought to be in an unstable state, and the ejection of the α, β, and γ rays leads to a more stable nucleus.

Most naturally occurring elements are not radioactive. All of the elements, however, have one or more isotopes that are radioactive; many of these are prepared synthetically today in nuclear reactors or in cyclotrons (several of the synthetic radioactive elements are listed in Table 1.1 and are identified by the superscript *d*). Since radioactive elements are easy to detect in very low concentrations because of the energetic α, β, or γ rays emitted, they are often used to "tag" molecules. For example, the fate of aspirin in the human body has been determined by tagging aspirin with a radioactive isotope of carbon, $_6C^{14}$. One way of doing this is to substitute $_6C^{14}$ for a small percentage of each of the $_6C^{12}$ atoms in aspirin, using the synthetic techniques devised by organic chemists. The excreta of the test subject are examined for radioactive compounds, which are separated and isolated. The multitude of nonradioactive compounds in the excreta can be safely ignored since they obviously do not contain any of the carbon atoms of the aspirin. The radioactive excretion compounds are then identified; in the case of aspirin they happen to be salicylic acid and gentisic acid, compounds that are closely related to aspirin. Once the products of metabolism of a drug are known, biochemists are in a better position to determine how the drug achieves its therapeutic action.

Very low concentrations of radioactive elements are used in these tagging experiments since radiation has a profoundly disruptive effect on the chemistry of living organisms, and general exposure to a large amount of radiation from radioactive elements (or from nuclear explosions!) can be fatal. However, in a technique used in cancer therapy, smaller amounts of radiation can be directed specifically to malignant tissue to destroy it without direct harm to the rest of the body.

THE PERIODIC TABLE

By the middle of the nineteenth century, it had been recognized that some of the elements have very similar properties. Lithium, sodium, and potassium, for example, are all highly reactive, soft metals. Chlorine, bromine, and iodine, on the other hand, make up a group of closely related nonmetals. Considerable effort was expended at that time to devise a general scheme to correlate the properties of the elements. The most useful arrangement of this type, the *periodic table*, was developed by the Russian chemist D. I. Mendeleev in 1869. In

Table 1.3   The periodic table[a]

The alkali metals
The alkaline earth metals
The halogens
The noble gases
Metalloids and nonmetals
Transition metals

PERIOD \ GROUP	I	II													III	IV	V	VI	VII	VIII
1	1 H Hydrogen																			2 He Helium
2	3 Li Lithium	4 Be Beryllium													5 B Boron	6 C Carbon	7 N Nitrogen	8 O Oxygen	9 F Fluorine	10 Ne Neon
3	11 Na Sodium	12 Mg Magnesium													13 Al Aluminum	14 Si Silicon	15 P Phosphorus	16 S Sulfur	17 Cl Chlorine	18 Ar Argon
4	19 K Potassium	20 Ca Calcium	21 Sc Scandium	22 Ti Titanium	23 V Vanadium	24 Cr Chromium	25 Mn Manganese	26 Fe Iron	27 Co Cobalt	28 Ni Nickel	29 Cu Copper	30 Zn Zinc			31 Ga Gallium	32 Ge Germanium	33 As Arsenic	34 Se Selenium	35 Br Bromine	36 Kr Krypton
5	37 Rb Rubidium	38 Sr Strontium	39 Y Yttrium	40 Zr Zirconium	41 Nb Niobium	42 Mo Molybdenum	43 Tc Technetium	44 Ru Ruthenium	45 Rh Rhodium	46 Pd Palladium	47 Ag Silver	48 Cd Cadmium			49 In Indium	50 Sn Tin	51 Sb Antimony	52 Te Tellurium	53 I Iodine	54 Xe Xenon
6	55 Cs Cesium	56 Ba Barium	57 La Lanthanum °	72 Hf Hafnium	73 Ta Tantalum	74 W Tungsten	75 Re Rhenium	76 Os Osmium	77 Ir Iridium	78 Pt Platinum	79 Au Gold	80 Hg Mercury			81 Tl Thallium	82 Pb Lead	83 Bi Bismuth	84 Po Polonium	85 At Astatine	86 Rn Radon
7	87 Fr Francium	88 Ra Radium	89 Ac Actinium ‡																	

° **Lanthanides (rare earth metals)**

58 Ce Cerium	59 Pr Praseodymium	60 Nd Neodymium	61 Pm Promethium	62 Sm Samarium	63 Eu Europium	64 Gd Gadolinium	65 Tb Terbium	66 Dy Dysprosium	67 Ho Holmium	68 Er Erbium	69 Tm Thulium	70 Yb Ytterbium	71 Lu Lutetium

‡ **Actinides**

90 Th Thorium	91 Pa Protactinium	92 U Uranium	93 Np Neptunium	94 Pu Plutonium	95 Am Americium	96 Cm Curium	97 Bk Berkelium	98 Cf Californium	99 Es Einsteinium	100 Fm Fermium	101 Md Mendelevium	102 No Nobelium	103 Lw Lawrencium

[a] Adapted from Ernest Grunwald and Russell H. Johnsen, *Atoms, Molecules, and Chemical Change* (Englewood Cliffs, N.J.: Prentice-Hall, 1960).

the periodic table, the elements are listed in order of their atomic numbers and arranged in such a way that similarities become apparent; a modern version is given in Table 1.3. The vertical columns, called *groups*, contain elements with similar properties, which usually vary in a systematic way from top to bottom. Group I, for example, contains highly reactive elements called the alkali metals. They all react violently with water, chlorine, and so on, and the rates of these reactions increase with increasing atomic number. The horizontal rows of the table, called *periods*, contain series of elements in which the chemical properties change in discrete steps. For example, in the third period, the first three elements (Na, Mg, and Al) are metals, and they appear in order of decreasing reactivity with a reagent such as water. The next four elements are nonmetals, and they are listed in order of increasing reactivity with a reagent such as sodium. The last element, argon, is unreactive and, as we shall see later, it represents the logical end of a period in the periodic table. This arrangement of the elements is not accidental; instead, as will become apparent, it is a direct consequence of the structures of the atoms.

The electrons of an atom are not distributed uniformly about the nucleus, nor are they distributed randomly. They occupy, in fact, certain fixed orbits (or shells) of different energies and these orbits, furthermore, contain certain fixed numbers of electrons. In Figure 1.1, for example, the third electron of lithium was not placed in the shell already containing two electrons, but it was placed in a second shell. This second shell can contain a maximum of eight electrons. Since the number of electrons in an atom equals the number of protons (the atomic number), we readily see from the periodic table (Table 1.3) that all the electrons of the elements up to neon will fit into the first two electron shells. Sodium, the next heavier element, contains one electron in a third shell.

Actually, all electron shells *beyond the first one* are further divided into subshells containing electrons that differ slightly in energy. The second shell, for example, is made up of one subshell containing a maximum of two electrons (designated an *s* subshell) and another subshell at a higher energy that can hold a maximum of six electrons (designated a *p* subshell). The third shell is made up of three subshells (called *s*, *p*, and *d*) containing a maximum of two, six, and ten electrons, respectively. A complete listing of the electron subshells, along with an indication of their relative energies, is given in Figure 1.3. The symbols for the subshells (*s*, *p*, *d*, *f*) were introduced by spectroscopists long ago to indicate the character of lines in the emission spectra of the elements (*sharp, principal, diffuse,* and *fundamental*).

The systematic development of the electron arrangements for the first 50 elements is given in Table 1.4. The filling of the subshells is straightforward up to argon (Ar). At this point, our simple order is interrupted; the 4s subshell is of

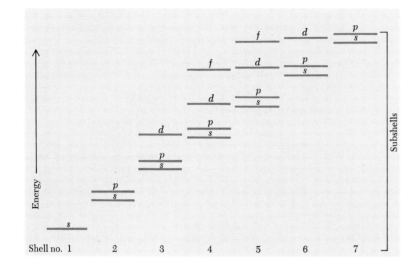

Shell no.   1   2   3   4   5   6   7

Figure 1.3   ***Relative energies*** *of the electron shells and subshells.*

lower energy than is the 3*d* subshell (Figure 1.3), and the next two electrons go into the fourth shell. The 3*d* subshell is filled next. Note that for elements with an incomplete 3*d* subshell, the electron distribution is irregular and either one or two electrons are found in the 4*s* subshell. The order in which the remaining subshells are filled is rather complex. A simple scheme for determining the general order is given in Figure 1.4 to aid the reader in outlining the electronic configurations of the remaining 53 heavier elements not listed in Table 1.4.

To use the chart, plot the levels of all the electrons in an atom, filling the sublevels in the order indicated by the arrows; start at the top of the figure and work downwards until the electrons are exhausted. The arrows in Figure 1.4 stop at the 5*f* level since atoms have not yet been detected or synthesized with electrons in higher energy sublevels. That is, the highest energy electrons of lawrencium ($_{103}$Lw) are in the 5*f* sublevel (note: *f* subshells contain a maximum of 14 electrons).

The electronic structures of the atoms (Table 1.4) and the positions of the elements in the periodic table are directly related. A correlation of Tables 1.3 and 1.4 will show, for example, that all the alkai metals (group I) contain a single electron in the outermost *s* subshell; further, all the halogens (group VII) contain five electrons in the outermost *p* subshell, and all the noble gases (group VIII) have a filled set of *s* and *p* subshells. In general, in each group of the periodic table, all the members have the same number and distribution of electrons in the outermost shell, and in each period, the number of electrons in this outermost shell increases in a regular way. The periodicity of the elements is therefore accounted for by the periodicity of the electronic structures.

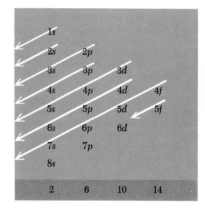

Figure 1.4   ***The electron subshells*** *listed in order of increasing energy.*

MATTER, ATOMS, AND MOLECULES

Table 1.4 *Electronic structures of the atoms*

ELEMENT	ATOMIC NO.	1s	2s	2p	3s	3p	3d	4s	4p	4d	4f	5s	5p	5d
		1	**2**		**3**			**4**				**5**		
H	1	1												
He	2	2												
Li	3	2	1											
Be	4	2	2											
B	5	2	2	1										
C	6	2	2	2										
N	7	2	2	3										
O	8	2	2	4										
F	9	2	2	5										
Ne	10	2	2	6										
Na	11				1									
Mg	12				2									
Al	13	Neon core			2	1								
Si	14	of			2	2								
P	15	10 electrons			2	3								
S	16				2	4								
Cl	17				2	5								
Ar	18				2	6								
K	19							1						
Ca	20							2						
Sc	21						1	2						
Ti	22						2	2						
V	23						3	2						
Cr	24						5	1						
Mn	25						5	2						
Fe	26				Argon core		6	2						
Co	27				of		7	2						
Ni	28				18 electrons		8	2						
Cu	29						10	1						
Zn	30						10	2						
Gd	31						10	2	1					
Ge	32						10	2	2					
As	33						10	2	3					
Se	34						10	2	4					
Br	35						10	2	5					
Kr	36						10	2	6					
Rb	37											1		
Sr	38											2		
Y	39									1		2		
Zr	40									2		2		
Cb	41									4		1		
Mo	42				Krypton core					5		1		
Tc	43				of					6		1		
Ru	44				36 electrons					7		1		
Rh	45									8		1		
Pd	46									10				
Ag	47									10		1		
Cd	48									10		2		
In	49									10		2	1	
Sn	50									10		2	2	

The outermost shell of electrons, called the *valence shell*, determines the chemical behavior of most of the elements with which we shall be concerned (however, in the transition metals and the rare earths, the partially filled *d* or *f* subshells of the next-to-outermost shell of electrons can also undergo changes in the number of electrons). Since the inner shells are not directly involved in chemical bonding (with the exception just noted), they are usually omitted from electron diagrams of the atom and the atoms are represented simply as shown here, with dots equal to the number of valence electrons:

·Li   :Be   :B·   :C: :N: :Ö: :F: :Ne:

Usually no attempt is made to distinguish the sublevels in these simple representations of the atom; their chief value lies in focusing attention on the total number of electrons in the valence shell. This number is of importance in determining the types of chemical reactions that an element undergoes.

In the 1920s, the French physicist Louis de Broglie proposed that electrons should show wavelike properties; that is, that a beam of electrons should have many of the properties of a beam of light. Within a few years, this view of the electron was verified experimentally, and today the electron microscope, which is based on this principle, may be found in most large research laboratories. Also in the 1920s, a model of the atom based on this concept of the wave nature of electrons was developed, in particular by the German physicist Erwin Schrödinger. Fundamental to the modern picture of the atom is the Schrödinger equa-

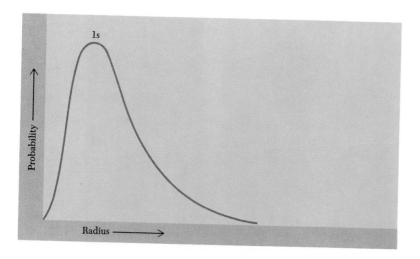

Figure 1.5 **Probability of finding a 1s electron** in a shell of unit thickness around the nucleus of an atom plotted against the radius of this shell.

tion, which describes the properties of the electron waves in atoms. Solutions to this equation are called wave functions, or *orbitals*, and they are a measure of the probability of finding an electron in a given region in space. That is, the definite, planar orbits of the Bohr (or planetary) atom are abandoned in favor of certain volumes in space in which the electrons move. The orbital that holds the 1s electrons is represented reasonably accurately by a sphere, for example. The probability distribution for a 1s electron as a function of the radius of this sphere is given in Figure 1.5. It can be seen from the diagram that the probability of finding the electron is greatest at a certain distance (which would correspond to the radius of the 1s electron shell in the planetary model of the atom) and that the probability drops off slowly, although it never vanishes. Strictly speaking, to account quantitatively for the one electron, the 1s orbital would have to be infinite in size. For representational purposes, however, a radius is usually chosen so there is a 90-percent probability of finding the electron within the volume determined by that radius.

The shapes of the 1s and 2p orbitals are represented in Figure 1.6. The p orbitals are shaped somewhat like dumbbells. There are three 2p orbitals, and these are oriented perpendicularly to one another. As we will see in the next few sections, this orientation provides an explanation for the 90° bond angles often found in compounds of the group V and group VI elements.

An orbital can be occupied by either one or two electrons. If two electrons are present, however, they must have opposite spins; the electrons are then said to be paired. The spin of an electron describes its behavior in a magnetic field; a simplified view is that the electron spins on its axis in either a clockwise or a counterclockwise fashion. The spin is usually represented by an arrow inside the symbol for an orbital (⬀ for one electron, and ⬀⬁ for paired electrons). We can now restate the electronic configurations of the atoms given in Table 1.4 in terms of the atomic orbitals involved (Table 1.5). It should be noted that in free atoms the electrons tend to remain unpaired in orbitals of the same subshell (for example, the p electrons in C, N, and O).

Figure 1.6   **Representations of the 1s and 2p orbitals.**

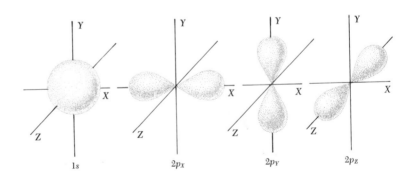

Table 1.5   *Electronic configurations of the second-period elements*

ELECTRON DISTRIBUTION

ELEMENT	1s	2s	2p		
Li	⇅	↑			
Be	⇅	⇅			
B	⇅	⇅	↑		
C	⇅	⇅	↑	↑	
N	⇅	⇅	↑	↑	↑
O	⇅	⇅	⇅	↑	↑
F	⇅	⇅	⇅	⇅	↑
Ne	⇅	⇅	⇅	⇅	⇅

As we shall see later, the modern picture of the atom gives us a much better understanding of matter than did the old planetary model. The usefulness of a theory is determined largely by what it can predict, and the new model of the atom is eminently successful in that respect.

It is easily verifiable that mixtures of many of the elements are unstable, and that reactions, often of a violent nature, occur when such mixtures are prepared. Furthermore, most of the other elements can be made to react with one another through the use of heat, light, or other forms of energy. The products of these reactions are called *compounds*. The properties of compounds are usually quite different from those of the constituent elements. For example, the reaction of the highly active, silvery metal sodium with the greenish gas chlorine yields sodium chloride (common table salt), a white, crystalline solid with a very high melting point.

Three elements in the periodic table have never been brought into chemical combination, however, despite intensive efforts to achieve this result. These elements are helium, neon, and argon, members of group VIII (the noble gases). The other elements in the group—krypton, xenon, and radon—are also inert, and compounds of these elements were not prepared until 1962, and then only under very special conditions. This reluctance of the noble gases to enter into chemical reactions indicates that the electronic configurations of these elements are particularly stable ones. Furthermore, it appears that the other elements react in such a way as to gain or approach the electron configuration of the nearest noble gas. The elements listed in the left-hand portion of the periodic table (those with fewer than four electrons in the valence shell) tend to reach noble-gas configurations by losing electrons, whereas those in the right-hand portion (those with more than four electrons in the valence shell) tend to reach those configurations by gaining electrons. For example, energy is required to remove one electron from a lithium atom:

$$\text{Li} \cdot + \text{energy} \longrightarrow \text{Li}^+ + 1\,\text{e}$$

Energy is set free, however, when a fluorine atom gains one electron:

$$:\overset{..}{\underset{..}{\text{F}}}\cdot + 1\,\text{e} \longrightarrow :\overset{..}{\underset{..}{\text{F}}}:^- + \text{energy}$$

Now if fluorine atoms are mixed with lithium atoms,

$$\text{Li} \cdot + :\overset{..}{\underset{..}{\text{F}}}\cdot \longrightarrow \text{Li}^+ :\overset{..}{\underset{..}{\text{F}}}:^- + \text{energy}$$

the compound lithium fluoride will be formed with the liberation of considerable energy. (A part of this energy comes from the attraction of the particles bearing opposite charges.) In this process, lithium has gained the helium electron configuration with a full, or closed, $1s$ shell, and fluorine has gained the neon electron configuration with closed $2s$ and $2p$ subshells.

When the neutral lithium atom loses one electron, a positively charged species results, thus bearing out the law of conservation of charge. When the neutral fluorine atom gains an electron, conversely, a negatively charged fluorine species is formed. Charged particles of this type are called *ions*, and compounds made up of ions, such as $\text{Li}^+ :\overset{..}{\underset{..}{\text{F}}}:^-$, are called *ionic compounds*. Ionic compounds other than this may be prepared by the reaction of the various metals (chiefly the elements in the left side of the periodic table) with the elements in group VII or with the lighter elements in groups V and VI. The formulas for most of the possible combinations can be determined from the electronic configurations listed in Table 1.4, that is, from the number of electrons that must be lost or gained to reach a noble-gas configuration. A number of compounds of this type are given below:

$$\text{Na}^+ :\overset{..}{\underset{..}{\text{F}}}:^- \qquad\qquad (\text{Cs}^+)_3 :\overset{..}{\underset{..}{\text{N}}}:^{3-}$$
Sodium fluoride   *Cesium nitride*

$$\text{K}^+ :\overset{..}{\underset{..}{\text{Cl}}}:^- \qquad\qquad \text{Ba}^{2+} :\overset{..}{\underset{..}{\text{O}}}:^{2-}$$
Potassium chloride   *Barium oxide*

$$(\text{Rb}^+)_2 :\overset{..}{\underset{..}{\text{O}}}:^{2-} \qquad\qquad \text{Sc}^{3+}(:\overset{..}{\underset{..}{\text{Br}}}:^-)_3$$
Rubidium oxide   *Scandium bromide*

Quite often, however, the ions in a compound only approximate a noble-gas configuration. This behavior is especially common among the ions of the transition metals and the rare earths (Table 1.3), although it is reasonably common in the elements of groups V and VI as well. A few examples are given:

$$K^+ . \overset{..}{\underset{..}{O}} : \overset{..}{\underset{..}{O}} : ^-$$
Potassium superoxide

$$Fe^{2+} (: \overset{..}{\underset{..}{Cl}} : ^-)_2$$
Ferrous chloride

$$Cu^+ : \overset{..}{\underset{..}{Br}} : ^-$$
Cuprous bromide

$$Fe^{3+} (: \overset{..}{\underset{..}{Cl}} : ^-)_3$$
Ferric chloride

Iron can combine with chlorine in two different ratios (as shown in the two formulas give above). This combining capacity of an element in an ionic compound is called its *valence;* it is represented normally by a number bearing a $+$ or $-$ sign to represent the charges on the ion. Thus, the elements listed above have the following valences: sodium ($+1$), fluorine (-1), potassium ($+1$), chlorine (-1), rubidium ($+1$), oxygen (-2; rarely -1), cesium ($+1$), nitrogen (-3; also more positive values in certain compounds), barium ($+2$), scandium ($+3$), iron ($+2$ and $+3$), and so on. In naming compounds of metals with a variable valence, the lower valence state is usually given the ending *-ous* and the upper valence state the ending *-ic;* for example, ferrous and ferric chlorides, cuprous and cupric bromides, and so on. The relationship of the valence of an element to the electronic structure of its atoms, and to the group number of the element in the periodic table, should be noted. The electronic structures of certain atoms and their ions are given in Table 1.6.

In addition to the simple ions that we have discussed, a number of complex ones, such as carbonate ion ($CO_3{}^{2-}$), sulfite ion ($SO_3{}^{2-}$), sulfate ion ($SO_4{}^{2-}$), nitrate ion ($NO_3{}^-$), and phosphate ion ($PO_4{}^{3-}$), exist (see page 36 for the electronic structures of the corresponding acids—that is, of the protonated

Table 1.6 **A comparison of the electronic structures of certain atoms and their ions**

ELEMENT	SUBSHELL						
	1s	2s	2p	3s	3p	3d	4s
Li	2	1					
Li$^+$	2						
F	2	2	5				
F$^-$	2	2	6				
Na	2	2	6	1			
Na$^+$	2	2	6				
Cl	2	2	6	2	5		
Cl$^-$	2	2	6	2	6		
Fe	2	2	6	2	6	6	2
Fe^{2+}	2	2	6	2	6	6	

neutral forms). Many ions are absolutely essential for life. For example, sodium and potassium ions are necessary for the operation of nerves, nitrite ion (NO_2^-) is involved in the conversion of NO_3^- ion to ammonia by bacteria, and PO_4^{3-} ions are involved in the digestion of foods (synthesis of ATP) and in the synthesis of DNA (the molecule in the cell that transmits hereditary information to the next generation).

Since the ions in an ionic compound bear opposite charges, they are attracted to one another by a considerable electrostatic force. This attractive force between the ions is referred to as an *ionic bond.* The charges on the ions are usually spherical in distribution and each positive ion can, therefore, attract several negative ions and each negative ion can attract several positive ions. The result is that in ionic compounds, the charged ions tend to form aggregates in which the ions alternate in the structure. A three-dimensional solid aggregate of this type with a regular structure is called a *crystal.* An example of this regular structure is found in the sodium chloride crystal, a representation of which is given in Figure 1.7.

Ionic compounds tend to be hard substances with high melting points. These properties are consistent with their crystalline structures. In order to cleave a crystal, or to melt it, a very large number of strong bonds must be broken; thus either a large physical force or a very high temperature is required (Table 1.8).

Figure 1.7   ***The arrangement of ions in a*** *crystal of sodium chloride.*

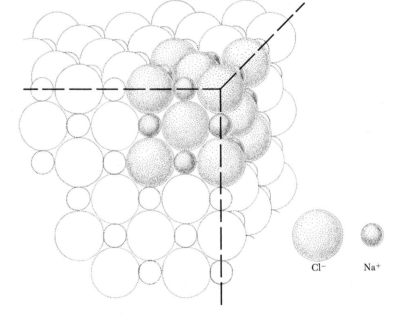

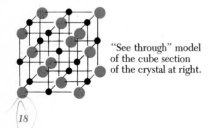

"See through" model of the cube section of the crystal at right.

Cl^- Na^+

Atoms can reach the noble-gas configuration not only through the gain or loss of electrons, but also through the sharing of electrons ($:\ddot{F}\cdot + :\ddot{F}\cdot \longrightarrow :\ddot{F}:\ddot{F}:$). This sharing of electrons usually occurs between identical atoms or between atoms found near the top right-hand part of the periodic table. In these cases, an electron donor is not reacting with an electron acceptor, a circumstance that is necessary to the formation of ionic compounds. The shared electrons circulate about both nuclei and they are treated as if they were in the valence shell of each atom. The attractive force of the two nuclei for the shared electrons holds the unit, or molecule, together (where we define *molecule* as an uncharged particle made up of two or more covalently bonded atoms).

Bond formation of this type accounts for the fact that most of the nonmetals exist in nature not in the atomic form, but in the molecular form. The atoms in molecules of the shared-electron type are said to be connected by *covalent bonds* and the substances so formed are called *covalent compounds*. The molecular forms of various nonmetals are given in Figure 1.8, along with their structures.

The *molecular weight* of a covalently bonded compound is defined as the sum of the atomic weights of all the atoms in the molecule. A second useful unit is the *gram molecular weight,* or *mole* for short, which is a weight in grams of the compound numerically equal to the molecular weight; such a quantity of matter contains 6.02×10^{23} molecules (Avogadro's number). Very often the term "gram molecular weight" is used when the weight aspect of this quantity is to be emphasized, and the term "mole" when the number aspect (that is, the number of particles) is the feature of interest. A summary of these and other chemical units is given in Table 1.7.

The absolute weight of a molecule may be found by dividing the gram molecular weight by 6.02×10^{23}, just as the absolute weight of an atom may also be found by dividing the gram atomic weight by Avogadro's number.

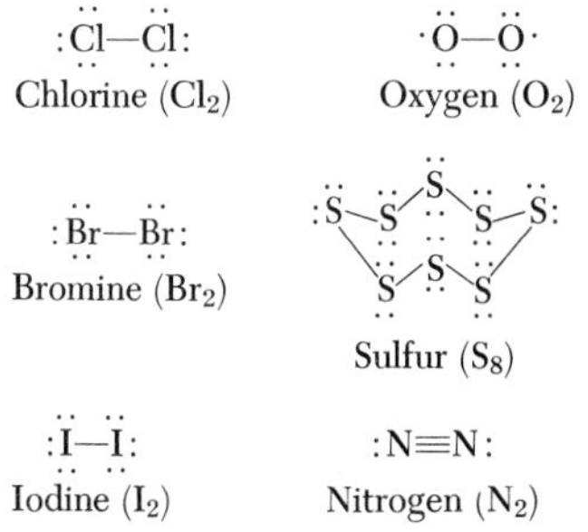

Covalent molecules may also be formed by the combination of different kinds of atoms, principally from the elements located in the right-hand portion of the periodic table. For example, 1 mole of nitrogen reacts with 3 moles of hydrogen at a high temperature to give 2 moles (34.061 g) of ammonia (NH_3).

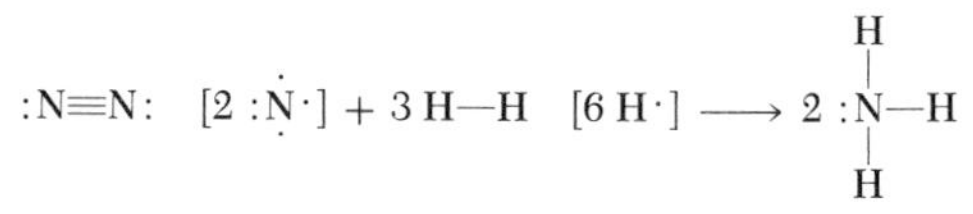

$$:N\equiv N: \quad [2\ :\dot{N}\cdot] + 3\ H-H \quad [6\ H\cdot] \longrightarrow 2\ :\overset{\displaystyle H}{\underset{\displaystyle H}{N}}-H$$

It is convenient to visualize the reaction in terms of the electronic structures of the atoms involved, as if the molecules break up into atoms first, and these react to give ammonia. We should recognize, however, that the reaction as it

:Cl—Cl:	·O—O·
Chlorine (Cl_2)	Oxygen (O_2)
:Br—Br:	
Bromine (Br_2)	Sulfur (S_8)
:I—I:	:N≡N:
Iodine (I_2)	Nitrogen (N_2)

Figure 1.8 **Molecular forms of some common elements** *in groups V, VI, and VII of the periodic table. The covalent bonds are represented by a dash (one dash for each electron pair involved).*

Table 1.7   **Summary of chemical units**

| | | ELEMENTS | | | COMPOUNDS |
| | | EXAMPLES | | | EXAMPLE |
QUANTITY	TERM	C	Cl	TERM	CCl_4
Smallest particle	*1 atom*	*1 average atom of C*	*1 average atom of Cl*	*1 molecule*	*1 average molecule of CCl_4*
Weight of particle relative to $_6C^{12}$ as <u>*12*</u> *AMU*	*atomic weight*	*12.01115 AMU*	*35.453 AMU*	*molecular weight*	*[12.01115 + 4(35.453)] 153.823 AMU*
6.02 × 10²³ particles (Avogadro's number)	*1 gram atomic weight or 1 g atom*	*12.01115 g of C*	*35.453 g of Cl*	*1 gram molecular weight or 1 mole*	*[1 g atom C + 4 g atoms Cl = 12.01115 g + 4 (35.453) g] 153.823 g of CCl_4*

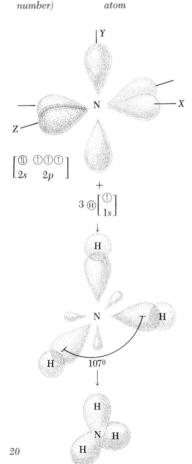

actually occurs may involve initially a direct interaction of the intact nitrogen molecules with hydrogen atoms.

The formation of covalent compounds can be profitably reconsidered from the standpoint of the atomic orbitals involved. For example, the formation of a hydrogen molecule from two hydrogen atoms can be represented as:

$$2 \, ⒣ \longrightarrow ⒣⒣ \longrightarrow (\text{H H})$$

The covalent bond is formed by the coalescing or overlapping of the $1s$ atomic orbitals of the hydrogen atoms. The new orbital is called a *molecular orbital* and it encompasses both nuclei in the molecule. There is a certain probability of finding the electron anywhere within the molecular orbital, although the probability is highest in the region between the nuclei; this is represented by a pulling in of the electron cloud from the far side of the atomic orbitals to the center of the molecular orbital. In a similar fashion, the reaction of hydrogen atoms with nitrogen atoms to form ammonia can be represented as shown at the left (from Table 1.5 and Figure 1.6).

Here, three molecular orbitals are formed by the overlapping of the $1s$ orbitals of hydrogen and the $2p$ orbitals of nitrogen; each molecular orbital can hold a maximum of two electrons (of opposite spin). For the sake of clarity, the $1s$ and $2s$ orbitals have been omitted from the figures of the nitrogen atoms since to a first approximation they are not involved in the bonding.

The length of a covalent bond is determined by a balance of forces: an attractive force resulting from the orbital overlap (the greater the overlap, the shorter and stronger the bond), and a repulsive force resulting from the charge interaction of the two positively charged nuclei of the atoms involved. This balance results in a N—H bond length of 1.01×10^{-8} cm in ammonia, and a H—H bond length of 0.74×10^{-8} cm in hydrogen. Molecular orbitals formed by the overlap of s and p atomic orbitals in any combination (for example, H_2 and NH molecular orbitals) are cylindrically symmetrical about the bond axes; such molecular orbitals are called σ (*sigma*) orbitals, and the bonds they represent are called σ bonds. A different type of molecular orbital, the π (*pi*) orbital, will be discussed in Chapter 3.

One might expect that the angles between the NH bonds in ammonia would be 90°, the value for the angles between the p orbitals of the nitrogen atom. But the experimental value, as determined by electron diffraction experiments, is 107° (all three HNH angles are the same). An explanation based on the polarity of the NH bonds has been advanced to account for this discrepancy. The electrons in a bond are shared equally by the two atoms only when those atoms are identical, as in H_2, F_2, and so on. Under these circumstances, the number of electrons around each atom exactly equals the number of protons in the nucleus. When different atoms are involved, one of the atoms forming the bond is likely to be more electronegative than the other. (The *electronegativity* of an atom is the power of attraction of that atom for the bonding electrons; see pp. 33–34.) In the case under discussion, the nitrogen atom is more electronegative than the hydrogen atom, with the result that in the NH bond the electrons are closer to the nitrogen nucleus than they are to the hydrogen nucleus. As a result, the nitrogen atom bears a slight negative charge and the hydrogen bears a slight positive charge (symbolized by δ^- and δ^+,° as in $^{\delta-}N—H^{\delta+}$). Bonds of this type are known as *polar bonds*, and they represent a halfway point between the symmetrical covalent bond (in H_2, F_2, and so on) and the ionic bond.

To return to the HNH bond angles in the ammonia molecule, it has been proposed that the abnormally large bond angles (107°) result from the mutual repulsions of the three partially charged hydrogen atoms in the molecule. Some support for this argument comes from a study of the bond angles in the hydrides of three other group V elements. If our reasoning concerning the bond angles in ammonia is correct, the hydrides of these heavier elements should have bond angles nearer the theoretical value of 90° since these elements—phosphorus, arsenic, and antimony—are less electronegative than is nitrogen, and since they have larger atomic radii (which would lead to greater hydrogen-hydrogen distances in the hydrides). The experimentally determined bond angles for these hydrides are indeed very close to the theoretical value; the experimental values are 93° for phosphine (PH_3), 92° for arsine (AsH_3), and 91° for stibine (SbH_3).

° The Greek letter *delta* (δ) is often used to signify a small amount or a small change.

The reaction of hydrogen with oxygen to form water (H—Ö—H) is similar to the hydrogen–nitrogen reaction outlined above and the details of the process are left to the reader as an exercise.

COVALENT COMPOUNDS OF CARBON   We might expect, in view of the distribution of valence electrons in the carbon atom ($2s$ ⑪ $2p$ ①①; see Table 1.5), that the derivatives of carbon would be divalent, where the *valence* of a covalently bonded atom is defined as the number of electron-pair bonds that it forms. In its stable compounds, however, carbon is predominantly tetravalent, and in compounds such as CH_4, CCl_4, and so on, all four bonds are identical. These facts concerning the compounds of carbon are accounted for by the modern theory of bonding in the following way.

The electronic configurations listed in Tables 1.4 and 1.5 are the configurations of lowest possible energy; they are called the *ground states* of the atoms. Most atoms in the gaseous phase at room temperature would exist in their ground-state electron configurations. If energy is supplied, however, in the form of either heat or light, new electronic configurations can arise. An atom with an electronic configuration different from that of the ground state is said to be in an *excited state*. As an example, in one of the excited states of the hydrogen atom, the electron is in a $2p$ orbital, whereas in the ground state, as we have seen from Table 1.4, it is in the $1s$ orbital. If energy is supplied to a carbon atom, one of the $2s$ electrons is "promoted" to a $2p$ orbital to give a carbon atom in an excited state. Thus:

$$\text{C(ground state, } 2s \text{ ⑪ } 2p \text{ ①①○)} \longrightarrow \text{C(excited state, } 2s \text{ ① } 2p \text{ ①①①)}$$

Four orbitals are now available and four bonds could in principle be formed with hydrogen. Considerable energy is released during the formation of a covalent bond, and the energy available from the formation of the two "extra" bonds to carbon would outweigh the energy required to promote a $2s$ electron to the $2p$ level. The explanation is still incomplete, however, because it would lead us to suspect that, in a molecule such as CH_4, three of the C—H bonds would be separated by 90° (the angle separating the p orbitals in the atom), and that the fourth C—H bond, involving the overlap of two s atomic orbitals, would have no particular angular dependence. We know, however, that all four bonds in CH_4 are identical and that the four bonds are separated by an angle of 109°28'. A second modification is required, and in this modification the $2s$ and the three $2p$ orbitals are combined to give a set of four equivalent orbitals, each with $\frac{1}{4}s$ character and $\frac{3}{4}p$ character:

$$2s \text{ ① } 2p \text{ ①①① } \longrightarrow sp^3 \text{ ①①①①}$$

The four equivalent sp^3 orbitals are then used in bond formation as shown at the left; in chemical terms,

$$\text{C} + 4\,\text{H} \longrightarrow CH_4$$

$C[sp^3$ ①①①①$] + 4\,H[s$ ①$]$

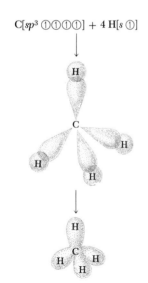

Orbitals formed by the combination of different types of orbitals are called *hybrid orbitals;* those of carbon that we have just discussed are known as sp^3 hybrid orbitals, in view of the type and number combined. The shape of an sp^3 orbital is shown in Figure 1.9. The sp^3 orbital is greatly concentrated in the bond direction and it forms stronger bonds than a p orbital. Theory predicts that four sp^3 orbitals should be directed to the corners of a tetrahedron and that sp^3 bonds should form an angle of 109°28′ (the value calculated from a regular tetrahedron). This is precisely the value that has been found by electron diffraction studies for the bond angles in CH_4 and related compounds. We shall make use of this bond angle in discussing the geometry of carbon compounds in Chapter 3.

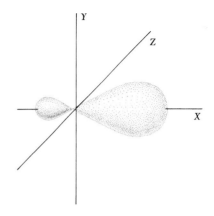

Figure 1.9   **An sp^3 orbital (on the X axis).**

THE PHYSICAL PROPERTIES OF COMPOUNDS   We have seen that the strong attraction of ions in ionic compounds accounts for the hardness of the latter compounds and their high melting points; in the next section, we will see why they dissolve only in polar solvents. Also, the molecules in polar covalent compounds show a moderate attraction for one another because of the charge separation, and as a result these compounds have moderate melting points and some solubility in both polar and nonpolar solvents. As might be expected by extrapolation, the molecules that comprise nonpolar covalent compounds show very little attraction for one another; therefore these compounds have low melting and boiling points. The molecules can mingle freely with other uncharged molecules, allowing nonpolar compounds to dissolve fairly well in nonpolar solvents. Conversely, nonpolar compounds are insoluble in polar solvents because the nonpolar molecules do not interact with the solvent molecules and they cannot break up the structure of the solvent. A general rule of solubility is "like dissolves like." A summary of these points is given in Table 1.8.

Table 1.8   **Physical properties of certain covalent and ionic compounds**

| | | | | | SOLUBILITY IN | |
COMPOUND	FORMULA	TYPE OF BONDING	MELTING POINT, °C	BOILING POINT, °C	POLAR LIQUID[a]	NONPOLAR LIQUID[b]
Hydrogen	H_2	Covalent nonpolar	−259	−253	—	—
Chlorine	Cl_2	Covalent nonpolar	−102	−35	—[c]	Soluble
Sulfur dichloride	SCl_2	Covalent polar	−78	59	—[c]	Soluble
Water	H_2O	Covalent polar and hydrogen bonds	0	100	Soluble	Insoluble
Beryllium chloride	$Be^{2+}(Cl^-)_2$	Ionic	440	520	Soluble	Insoluble
Lithium fluoride	Li^+F^-	Ionic	870	1670	Soluble	Insoluble

[a] Water (H_2O), for example.
[b] Carbon disulfide (CS_2) or hexane (C_6H_{14}), for example.
[c] The compound reacts with water and related solvents.

WATER   Water is by far the most common compound found in living organisms, and it performs many functions necessary to life as it has evolved on the earth. This presumably reflects the origin and early development of life in the sea. In any case, the high solvent ability of water and its roles in chemical reactions of the body, in maintaining a constant temperature in warm-blooded animals, and in lubricating the body tissues demonstrate that water is admirably suited to the needs of living organisms. These functions of water, furthermore, follow naturally from its structure. Water molecules (H_2O) contain polar covalent bonds between oxygen and hydrogen. Oxygen is the more electronegative of the pair (see p. 33; Table 2.2) and the following partial charges exist:

$$\delta-O \begin{matrix} H^{\delta+} \\ \\ H^{\delta+} \end{matrix}$$

A simpler formula $\quad ^-O \begin{matrix} H^+ \\ \\ H^+ \end{matrix} \quad$ and a molecular orbital picture

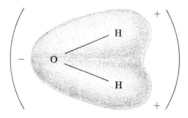

are also used to indicate the water molecule. This charge separation in the bonds accounts for both the unusual ability of water to dissolve ionic and polar compounds and the relatively high boiling point of water.

As was indicated previously, strong forces hold an ion in a crystal of an ionic compound. Bathing such a crystal with a nonpolar liquid such as hexane (page 58) or bromine does not cause it to dissolve since no source of energy exists to break the ion away from the rest of the crystal. Water, however, will dissolve most ionic compounds because the polar bonds in water enable it to react weakly with the ions, as shown at the top of page 25. Thus, the partial negative charge on oxygen leads to weak bonding with the positively charged sodium ion. The net energy gained by this weak interaction and that of the solvent with the chloride ion is enough to offset the energy required to remove the ion from the crystal. This clustering of solvent molecules about ions (and polar molecules) is called *solvation*.

The bond between water and the negative chloride ion is an example of an important class of weak bonds called *hydrogen bonds*. Since the hydrogen atom is very small, the chloride ion can approach to within a very short distance of it. At short distances, the electrostatic force of attraction of opposite charges

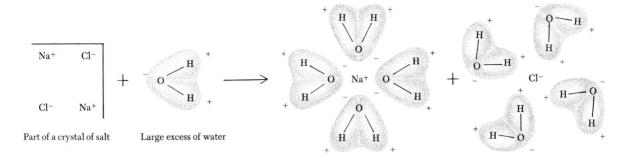

Na⁺	Cl⁻
Cl⁻	Na⁺

Part of a crystal of salt   Large excess of water

becomes fairly large; consequently, although the hydrogen atom is already bonded to oxygen, it can still bond weakly to chloride ion by the attraction of opposite charges. Hydrogen bonds of this type are about one-tenth as strong as is the average covalent bond.

The same phenomenon occurs in liquid water. In fact, extensive hydrogen bonding occurs between adjacent water molecules. Each water molecule has two hydrogen atoms to donate, and each oxygen atom in water can form two hydrogen bonds. This follows from the fact that the oxygen atom in water bears two pairs of unshared electrons; see previous section. Thus, a very complex network of hydrogen bonds exists in water, which we can only approximate in the planar drawing shown at the right.

The boiling point of water (100°) is relatively high (Table 1.8) because energy is required to break the hydrogen bonds in order to allow individual water molecules to reach the gaseous phase above the liquid.

Within the past decade, the pure deuterium isotope of hydrogen (Figure 1.2) has become readily available. On combustion, it forms deuterium oxide (D_2O), which, as might be expected, differs only slightly from ordinary water (H_2O). Bacteria and algae have been raised in pure D_2O and after a few generations fully deuterated organisms (those that have deuterium atoms replacing all the hydrogen atoms of the normal organisms) have arisen. Interestingly, the deuterated species often differ somewhat in shape from the normal organisms. Large quantities of the deuterated varieties are being raised, harvested, and degraded to give fully deuterated biological compounds which are of great value in research, but which would be extremely difficult to synthesize by ordinary organic laboratory techniques.

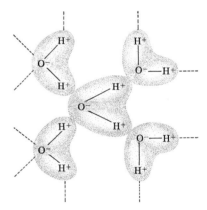

2 CHEMICAL REACTIONS

IN THE LAST CHAPTER WE POINTED OUT THAT MOST OF THE ELEMENTS ENTER into chemical combination to form various ionic and covalent compounds. In this chapter we shall be concerned with the further chemical reactions of these compounds, both with elements and with other compounds. A very large number of combinations is possible—in fact, literally millions of chemical reactions are known. Here, obviously, we can discuss only the more important of these reactions; the references given at the end of this volume should be consulted for a more complete discussion of others.

THE OXIDES

Virtually every element in the periodic table reacts with oxygen, and the products, called *oxides*, are of considerable theoretical and practical importance. The formulas for a few typical oxides are given in Table 2.1. A few of these oxides are widely distributed on earth: water requires no comment; carbon dioxide is found in the atmosphere and also in the form of calcium carbonate in marine shells, limestone, and marble; silicon dioxide is found in the pure state as quartz and sand, and in the bound state in most rocks.

Table 2.1   *Oxides of some of the more common elements*

Water (H_2O)
Hydrogen peroxide (H_2O_2)
Lithium oxide (Li_2O)
Beryllium oxide (BeO)
Boric oxide (B_2O_3)
Carbon monoxide (CO)
Carbon dioxide (CO_2)
Nitrous oxide (N_2O)

Nitric oxid (NO)
Nitrogen dioxide (NO_2)
Dinitrogen tetroxide (N_2O_4)
Fluorine oxide (F_2O)
Sodium oxide (Na_2O)
Magnesium oxide (MgO)
Aluminum oxide (Al_2O_3)

Silicon dioxide (SiO_2)
Phosphorus pentoxide (P_2O_5)
Sulfur dioxide (SO_2)
Sulfur trioxide (SO_3)
Chlorine monoxide (Cl_2O)
Titanium dioxide (TiO_2)
Zinc oxide (ZnO)

The oxides formed from the elements at the far left side of the periodic table are ionic compounds, and formulas such as $(Li^+)_2 : \ddot{O} : {}^{2-}$, for example, are satisfactorily accounted for by the principles of bonding developed in the last chapter. That is, each ion tends to gain the electronic structure of the nearest noble gas in the periodic table. Table 1.4 indicates how many electrons must be gained or lost to achieve that electronic structure. Since we will be dealing very largely with elements in the first three periods of the periodic table, we can use the simple rule that each ion, except for the hydrogen ion, tends to accumulate eight electrons in the outermost (valence) shell. This rule is often referred to as the *octet rule*. Hydrogen, the only member of period 1 that enters into chemical reactions, either gains one electron to fill its valence shell of two electrons or loses an electron to form a proton with no valence electrons.

The oxides formed from the elements at the upper right side of the periodic table are covalently bonded compounds, and their formulas and structures ($H : \ddot{O} : H$, $: \ddot{F} : \ddot{O} : \ddot{F} :$, and the like) are largely accounted for by similar principles. Each atom (other than hydrogen) again requires eight valence electrons. To determine the number of valence electrons an atom has in a covalent compound, add up the unshared electrons and then add the two electrons making up each bond to it. Thus, in H_2O ($H : \ddot{O} : H$) the valence shell of hydrogen contains two electrons and the valence shell of oxygen contains eight electrons. The electron pair of each bond is counted twice since the shared electrons circulate about each partner of the bond. By indicating all the electrons that the free atoms possess in the valence shell (page 13 and Table 1.5), the formulas of simple molecules can be readily obtained by inspection:

$$2 \; : \!\ddot{F}\! \cdot \; + \; 1 \; \cdot \!\ddot{O}\! \cdot \; \longrightarrow \; : \!\ddot{F} : \ddot{O} : \ddot{F}\! :$$

$$\cdot \dot{C} \cdot \; + \; 4 \; : \!\ddot{Cl}\! \cdot \; \longrightarrow \; \begin{matrix} : \!\ddot{Cl}\! : \\ : \!\ddot{Cl} : \!\dot{C}\! : \ddot{Cl}\! : \\ : \!\ddot{Cl}\! : \end{matrix}$$

CHEMICAL REACTIONS

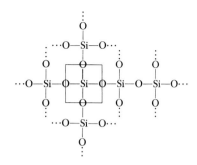

Certain other compounds, however (in particular, compounds of carbon, nitrogen, and oxygen), contain a different type of covalent bond called a *multiple bond*. In this type, two or three electron pairs are shared by the atoms involved. Examples are carbon monoxide, which contains a triple bond ($:C::O:$, or $:C\equiv O:$), carbon dioxide, which contains two double bonds ($:O=C=O:$), and dinitrogen tetroxide, which also has two double bonds (see the reaction at the end of the next paragraph). If only single bonds were present in these compounds ($:\ddot{C}:\ddot{O}:$, for example), the atoms involved would have electron structures quite far removed from those of the noble gases. Atoms of the elements in the second and third periods of the periodic table require eight valence electrons to reach the noble-gas configuration, and in the cases just cited, this octet of electrons is gained by multiple electron sharing. The nature of multiple bonds will be examined in more detail in the sections on ethylene and acetylene in Chapter 3.

Note that *pairs* of electrons normally are involved in covalent bonding; bonds made up of an odd number of electrons are quite rare. Some of the oxides of nitrogen and chlorine, however, have this rare structure. The molecules NO, NO_2, and ClO_2, for example, contain an odd number of electrons ($\cdot\ddot{N}::\ddot{O}:$, $:\ddot{O}:\dot{N}::\ddot{O}:$, and $:\ddot{O}:\dot{C}l:\ddot{O}:$). Molecules of this type are called *free radicals*. Free radicals have many properties in common with halogen atoms. For one thing, they tend to gain electrons in chemical reactions from elements at the left side of the periodic table: $Na\cdot + \cdot\ddot{N}::\ddot{O}: \longrightarrow Na^+(:\ddot{N}::\ddot{O}:)^-$. For another, at low temperatures they often double up to form a molecule with an even number of electrons:

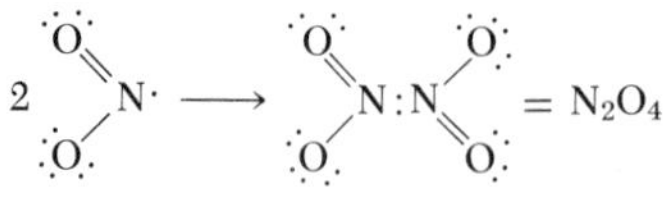

$$2\ \overset{\cdot\cdot}{\underset{:\ddot{O}:}{O}}N\cdot \longrightarrow \overset{\cdot\cdot}{\underset{:\ddot{O}:}{O}}N:N\overset{\ddot{O}\cdot}{\underset{\ddot{O}:}{}} = N_2O_4$$

Dinitrogen tetroxide

The oxides formed from the elements in the center part of the periodic table are unusual for quite other reasons. Boric oxide, aluminum oxide, silicon dioxide, and oxides of most of the other elements in this group exist as giant molecules containing many thousands of atoms. The silicon dioxide molecule, for example, consists of a very large network of alternating oxygen and silicon atoms (see Figure 2.1). The molecule is highly regular; it is made up of chains of a small "repeat unit," which in Figure 2.1 is enclosed in a rectangle. The four oxygen atoms surrounding each silicon atom are actually at the corners of a tetrahedron in which the silicon is centered; the geometry of each silicon atom thus resembles that of methane, CH_4 (see Chapters 1 and 3). The molecule in Figure 2.1 is an oversimplification, drawn in one plane for convenience.

Since each oxygen is shared by two silicon atoms (the rectangles shown bisect each oxygen), the repeat unit is SiO_2. The giant molecule, therefore, has a molecular formula of $(SiO_2)_n$, where n is some very large number. Such large molecules, usually with some simple repeat unit, are called *polymers;* other examples are covered in Chapters 3 and 6. For convenience, oxides of this type are referred to by their repeat unit; that is, silicon dioxide (quartz) $= SiO_2$, and aluminum oxide $= Al_2O_3$. As a result of their extremely high molecular weights, these oxides are completely insoluble in ordinary solvents and have very high melting points; they are used as insulation linings in furnaces, rockets, and other structures exposed to high temperatures.

The formulas we have been using represent *pure* compounds—that is, substances containing only a single kind of molecule. The ratio of carbon atoms to oxygen atoms in a large sample of carbon dioxide (CO_2) is the same, therefore, as the ratio in the individual molecules, namely 1/2. It also follows that the weight ratio of the two elements in the large sample is the same as the weight ratio in the individual molecules. Thus:

$$\text{Weight ratio of C to O in 1 molecule of } CO_2 = \frac{1 \text{ carbon atom} \times \dfrac{12 \text{ AMU of C}^\circ}{1 \text{ carbon atom}}}{2 \text{ oxygen atoms} \times \dfrac{16 \text{ AMU of O}}{1 \text{ oxygen atom}}} = \frac{12 \text{ AMU of C}}{32 \text{ AMU of O}} = \frac{1}{2.67} = \frac{0.375}{1}$$

$$= \frac{0.38 \text{ (AMU of C)}}{1 \text{ (AMU of O)}}$$

$$\text{Weight ratio of C to O in 1 mole of } CO_2 = \frac{1 \text{ gram atom of C} \times \dfrac{12 \text{ g of C}}{1 \text{ gram atom of C}}}{2 \text{ gram atoms of O} \times \dfrac{16 \text{ g of O}}{1 \text{ gram atom of O}}} = \frac{12 \text{ g of C}}{32 \text{ g of O}} = \frac{0.38 \text{ (g of C)}}{1 \text{ (g of O)}}$$

We see that the weight ratio of carbon to oxygen in 1 molecule of CO_2 is 0.38/1 and in 1 mole of CO_2 it is also 0.38/1. Thus, the weight ratio in any quantity of a pure compound can be calculated simply from the formula of the compound. Calculate the weight ratios of nitrogen to oxygen in N_2O, NO, NO_2 and N_2O_4. (*Answers:* 1.75/1, 0.875/1, 0.437/1, and 0.437/1, respectively.) Now divide the weights of N that reacted with one part of O by the lowest value (0.437). (*Answer:* 4, 2, 1, 1). This type of calculation, showing that one element reacts in integral ratios by weight with a fixed weight of another element (the law of multiple proportions), provided one of the foundations for the atomic theory of matter as it was developed in the nineteenth century.

° Note: $\dfrac{12 \text{ apples}}{1 \text{ dozen}}$ reads "12 apples per dozen" where the word *per* implies division.

Using a similar approach, we can calculate the percentage by weight of carbon in CO_2.

Percent of x in y = parts of x per 100 parts of y

$$= \frac{\text{parts } x \text{ by wt}}{1 \text{ part } y \text{ by wt}} \times 100 = \frac{\text{wt of } x}{\text{wt of } y} (100)$$

From the formula of carbon dioxide:

$$\text{Percentage of C in } CO_2 = \frac{12 \text{ g C}}{44 \text{ g } CO_2} (100) = 27.3$$

Percentage of O in CO_2 = 72.7

Carbon dioxide, therefore, is composed of 27.3 percent C and 72.7 percent O. Calculate the composition of NO, NO_2, and N_2O_4. (*Answers:* 46.7 percent N, 53.3 percent O; 30.4 percent N, 69.6 percent O; and 30.4 percent N, 69.6 percent O, respectively.)

The last calculation is often reversed, and the problem becomes one of determining the formula of a compound from the compositional data. For example, what is the formula of a compound that contains, by analysis, 30.4 percent N and 69.6 percent O? To solve this sort of problem, it is convenient to use an arbitrary amount of the compound (100 g, for example) and calculate the gram-atom ratio of the elements in that amount:

100 g of the compound contains 30.4 g N and 69.6 g O

$$100 \text{ g contains } \frac{30.4}{14.0} = 2.17 \text{ gram atoms of N}$$

$$100 \text{ g contains } \frac{69.6}{16.0} = 4.35 \text{ gram atoms of O}$$

The ratio is 2.17 to 4.35, or 0.499; within experimental error, this value is 0.5, and the ratio, therefore, is $\frac{1}{2}$ to 1. This value is also the ratio of the atoms in the molecule (remember that a gram atom of every element has the same number of atoms). Since a molecule cannot contain a fraction of an atom, the atom ratio is really 1 to 2, and the simplest formula we can have for this compound is N_1O_2, or NO_2.

In the absence of further information, we cannot tell whether the formula of our compound is NO_2, N_2O_4, N_3O_6, or some higher formula with the same *ratio* of nitrogen to oxygen. We need to know the *molecular weight* of the compound to decide this point. There are several experimental methods for getting approximate values of the molecular weights of compounds. Usually, we use solutions of the compounds in some liquid. Certain properties of the solutions, such as the melting point, boiling point, and vapor pressure, are found to depend on the number of molecules in solution (and for a fixed weight of compound this number depends on the molecular weight of the compound). If, by one of these methods, we learn that the molecular weight of our compound $(NO_2)_n$ is 92, then we know that $n = 2$, and the compound is N_2O_4. If molecular weight information is not available, only the simplest formula, NO_2, is reported; this simplest formula is called the *empirical formula* of the compound. Problem: Calculate the empirical formulas of compounds *A* and *B* from the analytical data given. Compound *A*: S, 40 percent; O, 60 percent. Compound *B*: P, 43.7 percent; O, 56.3 percent. (*Answers:* SO_3 and P_2O_5.)

A considerable amount of information concerning chemical reactions is summarized in a chemical equation, just as a considerable amount of mathematical information is summarized in an algebraic equation. Both types of equations must be balanced, however, before the information can be of use. The statement of fact that hydrogen reacts with oxygen to form water could be written $H_2 + O_2 \longrightarrow H_2O$, but the equation would not be balanced (the two sides are not equal). The right side tells us that two hydrogen atoms are present per oxygen atom. Two to one then must have been the ratio in which they reacted, and, therefore, this must also be the ratio of atoms on the left side of the equation. That is, four hydrogen atoms, or two hydrogen molecules, must have reacted with one oxygen molecule: $2\,H_2 + O_2 \longrightarrow H_2O$. The equation is still not complete, however, since two molecules of hydrogen and one molecule of oxygen yield not one but two molecules of water: $2\,H_2 + O_2 \longrightarrow 2\,H_2O$. The equation is now balanced; the two sides contain the same number of hydrogen atoms and the same number of oxygen atoms. Our equation is now consistent with the law of conservation of mass; that is, the total mass of the reactants equals the total mass of the products. We can now use the equation to calculate weight relations in chemical reactions. Our balanced equation tells us that: (1) hydrogen reacts with oxygen to form water; (2) two molecules of hydrogen react with one molecule of oxygen to form two molecules of water; and (3) 2 moles of hydrogen react with 1 mole of oxygen to give 2 moles of water. From our knowledge of atomic weights, molecular weights, and gram molecular weights, we can calculate, in addition, that 4.03 g of hydrogen react with exactly 32.00 g of oxygen to yield 36.03 g of water. These weight relations are very important in laboratory work since through their use we can readily calculate the quantity of one chemical needed to react completely with a given quantity of another. It must be stressed, however, that the calculations can be carried out only on balanced equations. The balancing of simple equations is generally a process of trial and error, as we have indicated in the hydrogen-oxygen example. The reader should balance the following equations:°

(*a*) $S_8 + O_2 \longrightarrow SO_2$
(*b*) $P_4 + O_2 \longrightarrow P_2O_3$
(*c*) $P_2O_5 + H_2O \longrightarrow H_3PO_4$

° *Answers:* The coefficients reading from left to right are (*a*) 1, 8, 8; (*b*) 1, 3, 2; and (*c*) 1, 3, 2.

Redox (from simplification of reduction-oxidation) reactions are chemical reactions involving both reduction and oxidation. "Oxidation" originally meant reaction with oxygen, but today, in a chemical context, it means the loss of electrons by a molecule or atom:

$$\text{Li·} \longrightarrow \text{Li}^+ + 1\,e$$

Reduction, on the other hand, means the gain of electrons by molecules or atoms:

$$:\overset{..}{\underset{..}{\text{Cl}}}\cdot + 1\,e \longrightarrow :\overset{..}{\underset{..}{\text{Cl}}}:^-$$

The balancing of redox reactions involving only simple ions is straightforward (the reactants and products must, of course, be known):

$$\text{Fe}^{3+} + \text{Sn}^{2+} \longrightarrow \text{Fe}^{2+} + \text{Sn}^{4+}$$

Simple inspection here indicates the proper coefficients needed for a balanced equation since each stannous ion (Sn^{2+}) loses two electrons (negative charges), whereas each ferric ion (Fe^{3+}) gains one electron. The balanced equation therefore reads:

$$2\,\text{Fe}^{3+} + \text{Sn}^{2+} \longrightarrow 2\,\text{Fe}^{2+} + \text{Sn}^{4+}$$

The reactions of the elements are, by and large, redox reactions and the balancing of these equations follows from the electron configurations of the elements and their ions. In the reaction of iron with chlorine, for example,

$$\text{Fe} + \text{Cl}_2 \longrightarrow \text{Fe}^{2+} + 2\,:\overset{..}{\underset{..}{\text{Cl}}}:^-$$

the iron atom has lost two electrons, and each chlorine atom has gained one electron. In this process the iron *becomes* oxidized, but it *acts* as a reducing agent (since it donates electrons); for chlorine, on the other hand, the situation is the opposite and it is reduced while *acting* as an oxidizing agent (since it gains electrons). Oxidation of one substance must necessarily accompany reduction of another, and the number of electrons lost by the reducing agent must equal the number of electrons gained by the oxidizing agent.

The product of the reaction of chlorine with an *excess* of iron is ferrous chloride (FeCl_2). If ferrous chloride is now treated with chlorine, the ferrous ion is oxidized to the ferric state:

$$2\,\text{Fe}^{2+}(\text{Cl}^-)_2 + \text{Cl}_2 \longrightarrow 2\,\text{Fe}^{3+}(\text{Cl}^-)_3$$

Many other oxidizing agents can convert ferrous ion into ferric ion; potassium permanganate is an example:

$$5\,Fe^{2+}(Cl^-)_2 + K^+MnO_4^- + 8\,HCl \longrightarrow$$
$$5\,Fe^{3+}(Cl^-)_3 + K^+Cl^- + Mn^{2+}(Cl^-)_2 + 4\,H_2O$$

We have seen in Chapter 1 that ionic compounds dissolve in water to give solvated ions, that is, ions surrounded by a shell of water molecules. For convenience, this shell is usually omitted, and a simple equation is used to describe the process:

$$\text{Solid } Fe^{3+}(Cl^-)_3 \longrightarrow Fe^{3+} + 3\,Cl^- \text{ (in solution)}$$

Once in solution, the ions essentially lead an independent existence and we can speak of free chloride ions, for instance. This interpretation accounts for the fact that a solution prepared by dissolving 1 mole of Li^+Cl^- and 1 mole of Na^+Br^- in 1 liter of water is identical to a solution prepared from 1 mole of Li^+Br^- and 1 mole of Na^+Cl^-.

With this in mind, we can rewrite our redox reaction to read:

$$5\,Fe^{2+} + K^+ + 8\,H^+ + 18\,Cl^- + MnO_4^- \longrightarrow$$
$$5\,Fe^{3+} + K^+ + Mn^{2+} + 18\,Cl^- + 4\,H_2O$$

Since the K^+ and Cl^- ions are not directly involved in the reactions, we can simplify the equation by subtracting these ions from both sides.

$$5\,Fe^{2+} + 8\,H^+ + MnO_4^- \longrightarrow 5\,Fe^{3+} + Mn^{2+} + 4\,H_2O$$

In this reaction, the ferrous ion is oxidized and the manganese atom is reduced. (Hydrogen and oxygen rarely change valence states in these reactions.) The last equation, as written, is balanced, both materially (both sides of the equation contain the same number of atoms) and electrically (the sum of the charges is the same on both sides of the equation); the electrical balance indicates that electrons are not created or destroyed, but are only transferred from one ion to the other.

Redox reactions involving complex ions and molecules are often difficult to balance by the trial-and-error method. As an aid to balancing these equations and also to understanding the redox principles, an *oxidation number* approach is often used. This is a number assigned to an atom or ion to reflect its stage of oxidation. By definition, the oxidation number of the free element is zero, and that of a simple ion is the valence of the particular ion. The assignment of oxidation numbers to the atoms in molecules or complex ions is somewhat arbitrary, on the other hand; the method is, in effect, an attempt to apply the simple ionic valence rules to molecules. Molecules are treated as if they contained only ionic bonds, and the electron pair of each bond is assigned to the more electronegative of the atoms making up the bond. The charge that remains on each atom is its oxidation number. In general, the electronegativities of the elements

Table 2.2 *Oxidation numbers*

ATOM OR ION	OXIDATION NUMBER
Cl *in* Cl_2	*0*
Cl^-	*−1*
Na^+	*+1*
H *in* H—O—H	*+1*
O *in* H—O—H	*−2*
N *in* NH_3	*−3*
C *in* O=C=O	*+4*
Mn *in* MnO_2	*+4*
Mn *in* MnO_4^-	*+7[a]*

$$^a MnO_4^- = \left[\begin{array}{c} :\overset{..}{O}: \\ | \\ :\overset{..}{O}-Mn-\overset{..}{O}: \\ | \\ :\overset{..}{O}: \end{array} \right]^-$$

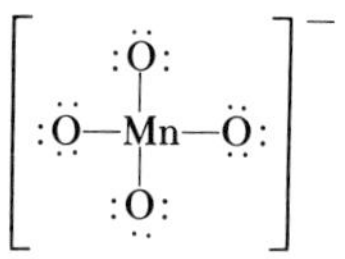

increase from left to right within any period in the periodic table, and from bottom to top in any group. Electronegativity is determined largely by the number of protons in the nucleus, the distance of the valence electrons from the positively charged nucleus, and the number of full electron shells between the nucleus and the valence electrons. Some examples of oxidation numbers are given in Table 2.2. It can be seen that the sum of the oxidation numbers of the atoms in a molecule is 0 and the sum in an ion is equal to the charge on that ion.

A change in the oxidation number of an atom implies a gain or loss of electrons, and the magnitude of this change tells us how many electrons were exchanged. We can demonstrate the value of oxidation numbers in balancing redox reactions by an example in which the permanganate ion (MnO_4^-) is involved.

Suppose that we are to balance the equation for the oxidation of ferrous ion in acidic solutions by permanganate ion. In the laboratory, this reaction could be carried out with ferrous chloride and potassium permanganate, with ferrous sulfate and sodium permanganate, or with some other combination of reagents. In any case, the products of the redox reaction are ferric ion and manganous ion (in oxidation reactions in acid solutions, the manganese atom from the permanganate ion usually ends up in the +2 valence state).

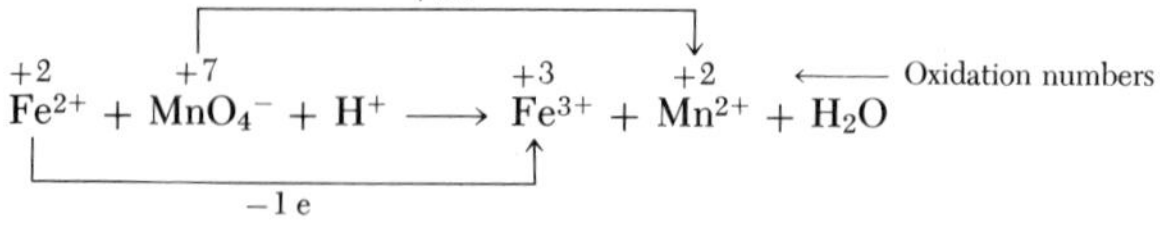

$$\overset{+2}{Fe^{2+}} + \overset{+7}{MnO_4^-} + H^+ \longrightarrow \overset{+3}{Fe^{3+}} + \overset{+2}{Mn^{2+}} + H_2O$$

+5 e (over MnO₄⁻ → Mn²⁺)
−1 e (under Fe²⁺ → Fe³⁺)
⟵ Oxidation numbers

We have indicated in the equation the oxidation numbers and also the number of electrons lost by each ferrous ion and the number gained by each manganese atom. It is obvious that five ferrous ions are oxidized by each permanganate ion:

$$5\,Fe^{2+} + MnO_4^- + H^+ \longrightarrow 5\,Fe^{3+} + Mn^{2+} + H_2O$$

The oxygen atoms in acid solutions appear either in the ion or as water; in our example four water molecules must be formed. Thus:

$$5\,Fe^{2+} + MnO_4^- + H^+ \longrightarrow 5\,Fe^{3+} + Mn^{2+} + 4\,H_2O$$

If the hydrogens are now balanced,

$$5\,Fe^{2+} + MnO_4^- + 8\,H^+ \longrightarrow 5\,Fe^{3+} + Mn^{2+} + 4\,H_2O$$

we find our equation balanced both materially and electrically.

The steps involved in balancing a redox reaction in basic solutions are the same with the exception that the "extra" oxygens in the complex ion must end up as hydroxide ion (OH^-), and occasionally water molecules must be added to the equation to balance the number of protons involved. The following redox reaction carried out in a basic solution should be balanced by the reader to gain a familiarity with the methods involved.°

$$CO + MnO_4^- + H_2O \longrightarrow CO_2 + MnO_2 + OH^-$$

Alchemists of the Middle Ages, in their attempts to classify matter, recognized two groups of compounds that were easy to characterize in terms of their properties. Members of the first group, called acids, have a sour taste, change the color of certain vegetable materials (indicators) in the same way, tend to dissolve metals, and react with members of the second group (bases). Bases have a brackish taste, turn indicators a different color, dissolve only a few special metals, and react with acids to give solutions that have neither acidic nor basic properties.

Later it was recognized that acids are compounds capable of furnishing hydrogen ions and bases are compounds capable of furnishing hydroxide ions.

$$HCl \longrightarrow H^+ + Cl^- \quad \text{(acid)}$$
$$NaOH \longrightarrow Na^+ + OH^- \quad \text{(base)}$$

Also, it was found that many of the common acids and bases can be prepared by the reaction of water with the oxides of the elements. Bases are prepared,

° *Answer:* The coefficients are 3, 2, 1, 3, 2, 2 in the order shown in the equation.

for example, from the oxides of elements found in the left side of the periodic table:

$$(Li^+)_2O^{2-} + H_2O \longrightarrow 2\,Li^+OH^-$$
Lithium hydroxide

$$Ca^{2+}O^{2-} + H_2O \longrightarrow Ca^{2+}(OH^-)_2$$
Calcium hydroxide

Very often, the same bases can be prepared from the reaction of the elements themselves with water; for example:

$$2\,Li + 2\,H_2O \longrightarrow 2\,Li^+OH^- + H_2$$
$$Ca + 2\,H_2O \longrightarrow Ca^{2+}(OH^-)_2 + H_2$$

Note that the reactions of the oxides with water are not redox reactions (no changes in the oxidation numbers occur), whereas the reactions of the elements with water are.

Acids are formed, in contrast, from the reaction of water with the oxides of the elements in the right side of the periodic table:

$$CO_2 + H_2O \longrightarrow H_2CO_3$$
Carbonic acid

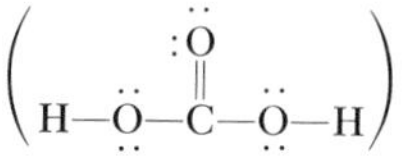

$$SO_2 + H_2O \longrightarrow H_2SO_3$$
Sulfurous acid

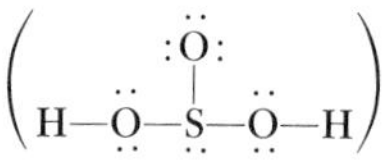

$$SO_3 + H_2O \longrightarrow H_2SO_4$$
Sulfuric acid

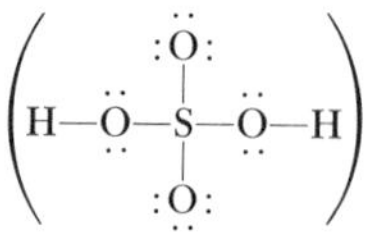

$$N_2O_5 + H_2O \longrightarrow 2\,HNO_3$$
Nitric acid

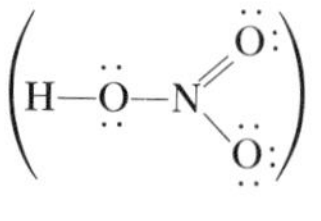

$$P_2O_5 + 3\,H_2O \longrightarrow 2\,H_3PO_4$$
Phosphoric acid

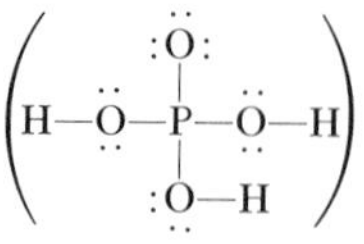

The oxides of the elements in the center part of the periodic table either are insoluble in water or yield hydroxides that have the properties of both weak acids and weak bases; these are called *amphoteric* compounds.

Structurally, H_2SO_4 is a hydroxide, and the question arises, why should

Ca(OH)$_2$ be a base and (HO)$_2$SO$_2$ be an acid? The answer has to do with the electronegativities of the calcium and sulfur atoms. Calcium has a low electronegativity, and even in the solid state of calcium hydroxide the calcium is fully ionized, Ca^{2+}(OH$^-$)$_2$. In contrast, the sulfur-oxygen bond is covalent. Since sulfur has a high electronegativity (see section on oxidation numbers), a large amount of energy would be required to ionize H$_2$SO$_4$ in the following sense:

$$\underset{\underset{\text{O}}{|}}{\overset{\overset{\text{O}}{|}}{\text{HO}-\text{S}-\text{OH}}} \longrightarrow \underset{\underset{\text{O}}{|}}{\overset{\overset{\text{O}}{|}}{\text{HO}^-{}^+\text{S}^+\text{OH}^-}}$$

Instead, the electronegativity of the sulfur augments the electronegativity of the oxygen atoms, and in pure liquid sulfuric acid, as a result, the OH bond is highly polar,

$$\underset{\underset{\text{O}}{|}}{\overset{\overset{\text{O}}{|}}{\overset{\delta^+\ \ \delta^-\quad\ \ \delta^-\ \ \delta^+}{\text{H}-\text{O}-\text{S}-\text{O}-\text{H}}}} \longrightarrow \text{H}^+\ \ {}^-\text{O}-\underset{\underset{\text{O}}{|}}{\overset{\overset{\text{O}}{|}}{\text{S}}}-\text{O}^-\ \ \text{H}^+$$

that is, the hydrogen atom has less than an equal share of the electron pair between hydrogen and oxygen. Compounds of this type tend to give up hydrogen ions readily, leaving the electron pair on the oxygen atom.

A second group of acids of considerable importance is made up of the halogen derivatives of hydrogen; these compounds—HF, HCl, HBr, and HI— are called the *hydrohalic* acids. They can be prepared from the elements (H$_2$ + Cl$_2 \longrightarrow$ 2 HCl), but a more convenient method for the three heaviest acids involves *hydrolysis* (reaction with water) of the phosphorus halides:

$$\text{PBr}_3 + 3\,\text{H}_2\text{O} \longrightarrow 3\,\text{HBr} + \text{H}_3\text{PO}_3$$

The PBr$_3$ can be easily prepared, in turn, by the reaction of phosphorus with bromine.

The compounds HF, HCl, HBr, and HI are gases at room temperature, whereas H$_2$SO$_4$ and HNO$_3$ are liquids. It is a surprising fact that the *pure* forms of these compounds do not have the properties that we associate with acids. These properties develop only when the compounds are dissolved in water or in some other very polar solvent. The dissolution is accompanied by the evolution of considerable heat, which is one of the indications that a chemical reaction has occurred. This chemical reaction is in the present time the transfer of a proton from the acid to the solvent:

$$\text{HNO}_3 + \text{H}_2\ddot{\text{O}}: \longrightarrow \underset{\underset{\text{H}}{|}}{\text{H}-\overset{\cdots}{\text{O}}{}^\pm\text{H}} + \text{NO}_3{}^-$$

$$\text{HCl} + \text{H}_2\text{O} \longrightarrow \text{H}_3\text{O}^+ + \text{Cl}^-$$

$$\text{H}_2\text{SO}_4 + \text{H}_2\text{O} \longrightarrow \text{H}_3\text{O}^+ + \text{HSO}_4{}^-$$

Table 2.3   **Typical salts**

NaI	*Sodium iodide*	$CaSO_4$	*Calcium sulfate*
$Ca(NO_3)_2$	*Calcium nitrate*	$Fe_2(SO_4)_3$	*Ferric sulfate*
$Al(NO_3)_3$	*Aluminum nitrate*	$NaHSO_4$	*Sodium bisulfate (or sodium hydrogen sulfate)*
Na_2CO_3	*Sodium carbonate*	K_3PO_4	*Potassium phosphate*
$KHCO_3$	*Potassium bicarbonate*	KH_2PO_4	*Potassium dihydrogen phosphate*
Li_2SO_3	*Lithium sulfite*		

At the halfway point of this proton transfer, a hydrogen-bonded species (Chapter 1) is formed ($H_2O\cdots H{-}Cl$, for example), but for HCl, H_2SO_4, and other strong acids, full transfer of a proton ultimately occurs to give a protonated water molecule, H_3O^+, called a *hydronium ion*. It is the *hydronium ions* that give the water solutions of these compounds "acidic" properties. The negative ions are not directly involved in reactions of acid solutions, as is shown by the fact that approximately the same amount of heat is liberated when we treat the solutions of any one of the hydrohalic acids with sodium hydroxide; this is illustrated in the following equation:

$$Na^+ + OH^- + H_3O^+(Cl^-, Br^-, I^-) \longrightarrow 2\,H_2O + Na^+(Cl^-, Br^-, I^-) + heat$$

One characteristic of acids and bases is their ability to neutralize one another. For example, 1 mole of any one of the hydrohalic acids will react with exactly 1 mole of any one of the group I hydroxides:

$$H_3O^+Cl^- + Na^+OH^- \longrightarrow 2\,H_2O + Na^+Cl^-$$

The product in the example cited is a water solution that has neither acidic nor basic properties; it is called a neutral solution. The solution contains, in fact, only sodium chloride (common table salt). In general, the reactions of acids with bases yield ionic compounds (often called salts) and water, as shown in the following equations:

$$H_3O^+NO_3^- + Li^+OH^- \longrightarrow Li^+NO_3^- + 2\,H_2O$$
$$(H_3O^+)_2SO_4^{2-} + 2\,K^+OH^- \longrightarrow (K^+)_2SO_4^{2-} + 4\,H_2O$$

Examples of salts prepared in this way are given in Table 2.3. The ionic charges have been omitted from the formulas for convenience.

In salt nomenclature, the name of the metallic element is first, followed by a name for the negative ion based on the name of the corresponding acid (page 36); the acid name is usually modified to end in *-ate* or *-ide*, although when there are two valence states, the names given to compounds of the lower valence state end in *-ite* (SO_3^{2-}, for example).

MOLARITY   It is often more convenient to use water solutions of acids and bases than the pure compounds. To do this conveniently, we employ a new unit, *molarity* (M), which is defined as the number of moles of a compound dissolved in 1 liter of solution. For example, 1 liter of a 1 molar (1 M) solution of nitric acid is prepared by adding water to 1 mole of nitric acid (63 g) until the volume of the solution reaches 1 liter. (This solution is slightly different in concentration from a solution prepared by adding 1 mole of nitric acid to 1 liter of pure water.) It follows from the definition that the product of the molarity of a solution and the volume of that solution (V) is equal to the number of moles of reagent in the system: $M \times V =$ moles. This interconversion of units can be demonstrated readily with the aid of a balanced equation:

$$H_2SO_4 \quad + \quad 2KOH \quad \longrightarrow \quad K_2SO_4 \quad + \quad 2\,H_2O$$

(a)   1 mole   +   2 moles $\longrightarrow$ 1 mole   +   2 moles
(b) 1 liter of 1 M H$_2$SO$_4$ soln. + 1 liter of 2 M KOH soln. $\longrightarrow$ 1 liter of 1 M K$_2$SO$_4$ soln. + 2 moles H$_2$O
(c) 1 liter of 1 M H$_2$SO$_4$ soln. + 2 liters of 1 M KOH soln. $\longrightarrow$ 1 liter of 1 M K$_2$SO$_4$ soln. + 2 moles H$_2$O

Each experiment [(a), (b), or (c)] will yield exactly 1 mole of K$_2$SO$_4$.

NORMALITY   One liter of 1 M H$_2$SO$_4$ contains twice as many hydrogen ions as does 1 liter of 1 M HCl. We often find it convenient to deal with solutions containing the same number of hydrogen ions or hydroxide ions per liter. For this purpose we make use of another unit, *normality* (N). The normality of an acid solution is the number of moles of hydrogen ion present per liter of solution, and the normality of a base solution is the number of moles of hydroxide ion per liter. That is, a solution containing 1 mole of KOH per liter is a 1 normal (1 N) solution—as well as a 1 M solution. A solution that contains 1 mole of H$_2$SO$_4$ per liter, however, is a 2 N solution—since two moles of hydronium ion are supplied by each mole of sulfuric acid. It follows then that 1 liter of any 1 N acid solution will exactly neutralize 1 liter of any 1 N base solution. The *equivalent weight* of an acid is that weight of the substance which furnishes 1 mole of H$_3$O$^+$; of a base, that weight which furnishes 1 mole of OH$^-$. Thus, the equivalent weight of H$_2$SO$_4$ is 98/2 = 49 g, the equivalent weight of NaOH is 40/1 = 40 g, and the equivalent weight of H$_3$PO$_4$ is 98/3 = 32.7 g. The general formula for this calculation is

$$\text{equivalent weight} = \frac{\text{molecular weight}}{\text{number of H or OH per molecule}}$$

One equivalent weight of any acid or base diluted to 1 liter final volume with water yields a 1 N solution of the acid or base. The normality of a solution, therefore, may be defined as the number of equivalent weights in a liter of that solution (eq wts/liter). A convenient formula to use for calculating the amount of one solution required to neutralize another is: $N_{acid} \times$ volume of acid solution $= N_{base} \times$ volume of base solution. From our definition of normality, the equation states, in effect: Number of equivalents of acid = number of equivalents of base. It follows from these equations that 1 liter of 5 N HCl will exactly neutralize 2 liters of 2.5 N NaOH, for example.

If all the ions that are not involved directly in a neutralization reaction are subtracted from both sides of the equation, a very simple equation results, as shown here:

$$H_3O^+Br^- + K^+OH^- \longrightarrow 2\,H_2O + K^+Br^-$$
$$H_3O^+ + OH^- \longrightarrow 2\,H_2O$$

The important reaction in neutralization, then, is the proton transfer from a hydronium ion to a hydroxide ion. This simplified view of neutralization led the Danish chemist J. N. Brønsted to propose more general definitions of acids and bases in 1923. According to the *Brønsted definitions*, acids are proton donors and bases are proton acceptors. This definition is in better accord with chemical facts; for example, it makes clear why pure ammonia (NH_3) is a base even though it is not a hydroxide. Thus:

$$H_3\ddot{O}^+\!:\!\ddot{Br}\!:^- + :NH_3 \longrightarrow NH_4^+\!:\!\ddot{Br}\!:^- + H_2\ddot{O}\!:$$

$$\text{or } H_3\ddot{O}^+ + :NH_3 \longrightarrow NH_4^+ + H_2\ddot{O}\!:$$

To return to our simple equation for acid-base reactions, we might suppose that pure water itself would contain small amounts of hydronium and hydroxide ions since water can act as both an acid and a base. Thus:

$$2\,H_2\ddot{O}\!: \longrightarrow H_3\ddot{O}\!:^+ + :\ddot{O}H^-$$

Measurements of the conductance of pure water show this to be true, although the extent of dissociation is low; the concentrations of H_3O^+ and OH^- in pure water at $25°C$ are both equal to 10^{-7} M. The product of these concentrations is found to be a constant at any given temperature; the value of this constant at $25°$ is 10^{-14}. We can express this as $[H_3O^+][OH^-] = 10^{-14}$. The brackets in the expression signify concentrations of hydronium and hydroxide ions expressed in molarity. This equation indicates that water solutions never contain only H_3O^+ ions or only OH^- ions, but always contain both. The relative proportions of the two determine the acidity or basicity of the solution. If we know one of the values, we are able to readily calculate the other by using the equation.

Suppose that a 0.1 M solution of HNO_3 is prepared by diluting 6.3 g of nitric

Table 2.4 **Measures of acidity**

[H$_3$O]$^+$, MOLES PER LITER	[OH$^-$], MOLES PER LITER	log [H$_3$O$^+$]	$-$log [H$_3$O$^+$] = pH
1×10^0	1×10^{-14}	0	0
1×10^{-1}	1×10^{-13}	-1	1
1×10^{-4}	1×10^{-10}	-4	4
1×10^{-7}	1×10^{-7}	-7	7
1×10^{-10}	1×10^{-4}	-10	10
1×10^{-13}	1×10^{-1}	-13	13
1×10^{-14}	1×10^0	-14	14

acid to 1 liter of water. The HNO$_3$ generates 0.1 mole of hydronium ion; this swamps the very small amount present in pure water (0.0000001 mole), and the final concentration of hydronium ion is, for all practical purposes, 0.1 M. If the final concentration of H$_3$O$^+$ is 0.1 M, then from the equation the final concentration of OH$^-$ must be 10^{-13} M:

$$[OH^-] = \frac{10^{-14}}{[H_3O^+]} = \frac{10^{-14}}{10^{-1}} = 10^{-13}$$

The ratio of the hydronium to the hydroxide ion concentration is extremely large ([H$_3$O$^+$]/[OH$^-$] = 10^{12}), and for most purposes only the acidic properties of such a solution need be considered.

In a graded series of concentrations—for example, 10^{-4} M H$_3$O$^+$NO$_3^-$, 10^{-3} M H$_3$O$^+$NO$_3^-$, and 10^{-2} M H$_3$O$^+$NO$_3^-$—the hydronium ion concentrations are a measure of the acidity of the solutions. These numbers are cumbersome to use, and therefore another unit of acidity, the *pH* unit, has been devised. The pH of a solution is the negative logarithm° of the hydronium ion concentration: pH $= -$log [H$_3$O$^+$]. For our three nitric acid solutions, the log [H$_3$O$^+$] would be -4, -3, and -2, and the pH would be 4, 3, and 2, respectively; that is, the smaller the pH value, the greater the acidity. More detailed examples of the relationships between the hydroxide ion concentration, the hydronium ion concentration, and the pH of acid and base solutions are given in Table 2.4. More complex examples are handled in a similar way:

$$[H_3O^+] = 5 \times 10^{-3}$$
$$pH = -\log 5 \times 10^{-3} = -1(\log 5 \times 10^{-3}) = -1(0.70 - 3) = 2.30$$

The use of pH units allows one to refer conveniently, but quantitatively, to the acidity of a solution; acidic solutions have a pH of less than 7, and basic solutions have a pH greater than 7.

° See Appendix B.

The acids and bases used as examples up to this point are essentially completely dissociated in solution; they are called *strong acids* and *bases*. Certain other acids and bases are only partially dissociated in solution; they are called *weak acids* and *bases*. Acetic acid (CH_3CO_2H, which we will symbolize as HOAc) is a typical weak acid. A 0.1 M solution of HOAc does not have a pH of 1 (H^+ concentration $= 10^{-1}$), but instead has a pH of about 3 (H^+ concentration $= 10^{-3}$). Since the hydronium ion concentration is low, we conclude that most of the acetic acid is present in solution in the undissociated form. We can indicate this state by an equation with double arrows to show that the species represented on both sides of the equation are present in solution at the same time, as follows:

$$HOAc + H_2O \rightleftharpoons H_3O^+ + OAc^-$$

When the acetic acid is first added to water, dissociation occurs to give hydronium and acetate ions. Acetate ion (OAc^-) happens to be a moderately strong base (by the Brønsted definition) and it reacts with the hydronium ion to give back acetic acid:

$$OAc^- + H_3O^+ \longrightarrow HOAc + H_2O$$

As the acetic acid dissociates (actually only a fraction of a second may be required for this process) the concentrations of H_3O^+ and OAc^- increase, and therefore the rate of the reverse reaction increases. A point is reached at which the rate of the forward reaction (number of HOAc molecules dissociating per second) equals the rate of the reverse reaction (number of H_3O^+ and OAc^- ions reacting per second) and at this point, the system is said to be at *equilibrium* (represented by the double-arrow equation):

$$HOAc + H_2O \rightleftharpoons H_3O^+ + OAc^-$$

A system in equilibrium shows no change in properties with time. The equilibrium ratio of products to reactants is fixed at a given temperature; this is expressed by the following equation:

$$\left[\frac{\text{product of the concentrations of the products}}{\text{product of the concentrations of the reactants}} \right] = \text{a constant}$$

Or, specifically for acetic acid,

$$\frac{[H_3O^+][OAc^-]}{[HOAc][H_2O]} = K'$$

where the brackets represent concentrations of the species at equilibrium, and

Table 2.5 **Concentrations (molar) in the acetic acid equilibrium**

COMPONENT	IMMEDIATELY AFTER MIXING AND BEFORE DISSOCIATION OCCURS	AT EQUILIBRIUM	
HOAc	0.1	0.1 − X	
H_3O^+	10^{-7} { *The value in pure water*	X	{ *The trace of H_3O^+ coming from the dissociation of water is neglected*
OAc⁻	0	X	

K' is called the equilibrium constant. In a dilute solution in water, the concentration of water $(55.6\ M)^°$ does not change appreciably during a chemical reaction. For convenience, then, the ionization equation is usually rearranged so that the water concentration is part of the constant.

$$\frac{[H_3O^+][OAc^-]}{[HOAc]} = K'[H_2O] = K$$

The dissociation constant (K) of an acid is a measure of the acidity of that acid. The values of K for strong acids such as HCl and H_2SO_4 are considerably greater than 1 (our previous assumption that these compounds are fully ionized in water was valid). The value of K for acetic acid (1.75×10^{-5}) is considerably less than 1. Other weak acids are carbonic acid $(H_2CO_3, K = 4.7 \times 10^{-7})$, nitrous acid $(HNO_2,\ K = 4.1 \times 10^{-4})$, and hydrogen sulfide $(H_2S, K = 1.1 \times 10^{-7})$.

Dissociation constants are easily calculated once the concentrations of the species present at equilibrium are known. For example, the following concentrations have been measured for a certain solution of nitrous acid at equilibrium: $[HNO_2] = 0.98\ M$; $[H_3O^+] = 0.02\ M$; and $[NO_2^-] = 0.02\ M$. If we substitute these values in our equilibrium expression,

$$K = \frac{[H_3O^+][NO_2^-]}{[HNO_2]} = \frac{(0.02)(0.02)}{0.98}$$

and perform the calculation, we get a dissociation constant for nitrous acid of 4.1×10^{-4}.

Once the dissociation constant is known, we can calculate the acidity of solutions of the acid. For example, suppose that we needed to know the hydronium ion concentration in a 0.1 M solution of acetic acid. The concentrations of the species present at equilibrium can be set up in terms of a single un-

---

$^°$ One liter of solution contains *approximately* 1,000 g H_2O = 1,000/18 = 55.6 moles H_2O (the molecular weight of H_2O is 18).

known quantity X, as shown in Table 2.5, where X = the amount of acid that has dissociated. The substitution of these values in our equilibrium expression gives us $X^2/(0.1 - X) = K$ ($K = 1.75 \times 10^{-5}$; see p. 43). In calculations for weak acids, X is normally dropped from the denominator since it is usually a relatively small number; that is, $0.1 - X$ is very nearly equal to 0.1. The equation then becomes $X^2/0.1 = 1.75 \times 10^{-5}$, and solving for X ($X^2 = 1.75 \times 10^{-6}$; $X = \sqrt{1.75 \times 10^{-6}} = \sqrt{1.75} \times 10^{-3} = 1.32 \times 10^{-3}$) gives us $1.32 \times 10^{-3} M$ as the concentration of hydronium ion in this solution at equilibrium. The percentage of dissociation of acetic acid in a 0.1 M solution can then be calculated:

$$\text{percentage of dissociation} = \left(\frac{1.32 \times 10^{-3}}{0.1}\right) 100 = 1.32\%$$

Only 1.32 percent of the acetic acid, then, is ionized in a 0.1 M solution at 25°C (room temperature).

INDICATORS

Very often it is necessary to determine the pH of a solution without knowing the amounts and kinds of acids present. A class of compounds called *indicators* is useful in assays of this type. Indicators are complex organic compounds (weak acids [HIn] and weak bases [In]) that have different colors, depending on whether they are in the ionized or nonionized form.

Phenolphthalein is an indicator that is colorless in acidic solutions and red in basic solutions:

$$\text{H In} + \text{OH}^- \longrightarrow \text{In}^- + \text{H}_2\text{O}$$
$$\quad \textit{Colorless} \qquad\qquad \textit{Red}$$

This change in color occurs at a pH very near 7. If a base is added slowly to a solution containing an acid and a small amount of phenolphthalein, the neutralization point will be reached at the instant the solution becomes red in color. This process of quantitative neutralization is called *titration*. The value of indicators in determining the neutralization point is apparent.

Other indicators exist that give colors in the entire range of the visible spectrum. The pH at which an indicator changes color is determined by its chemical makeup; enough indicators are available so that one can be chosen to reveal a change anywhere on the pH scale.

The materials responsible for the colors of most fruits and vegetables are indicators, and the color changes attendant on ripening result from the general lowering of acidity. Vegetable extracts, such as litmus and extracts of red cabbage, have been used as indicators in chemistry since the days of the alchemists.

The pH of mammalian blood is maintained at a value very close to 7.35. If the pH shifts by as small an amount as 0.2 of a unit, serious impairment of the functioning of the organism, or even death, may occur. Since the addition of as little as 10^{-6} mole (about 0.0004 g) HCl to 1 liter of water changes the pH by a full unit (from 7 to 6), it is apparent that living organisms must have some means of protecting themselves from sudden changes in acidity. The systems used to achieve this result are called *buffers*.

Buffers are mixtures of a weak acid and its salt or a weak base and its salt. We can illustrate the action of buffers with the aid of the expression for the equilibrium constant of acetic acid developed earlier.

$$HOAc + H_2O \rightleftharpoons H_3O^+ + OAc^- \qquad K = \frac{[H_3O^+][OAc^-]}{[HOAc]}$$

We can rearrange the equation:

$$K[HOAc] = [H_3O^+][OAc^-] \qquad \text{or} \qquad [H_3O^+] = K\frac{[HOAc]}{[OAc^-]}$$

and take logarithms° of each side:

$$\log[H_3O^+] = \log K + \log\frac{[HOAc]}{[OAc^-]}$$

Multiplying each side of the equation by -1 gives us:

$$-\log[H_3O^+] = -\log K - \log\frac{[HOAc]}{[OAc^-]} = -\log K + \log\frac{[OAc^-]}{[HOAc]}$$

Given our definition of pH, and the further definition that $-\log K = pK$, our expression then becomes:

$$pH = pK + \log\frac{[OAc^-]}{[HOAc]}$$

Suppose we now add sodium acetate (Na^+OAc^-) to this solution of acetic acid. Salts are ionized in solution and the acetate ion added is indistinguishable from the acetate ion formed by the ionization of acetic acid. Furthermore, if a large amount of sodium acetate is added, the amount of acetate in our expression above can be set equal to the concentration of the sodium acetate added since only a very small, and negligible, amount of acetate ion is formed by the dissociation of acetic acid itself. The expression then becomes:

$$pH = pK + \log\frac{\text{salt concentration}}{\text{acid concentration}} = pK + \log\frac{[OAc^-]}{[HOAc]}$$

° See Appendix B.

Since K, and therefore pK, is a constant, the expression tells us that the pH of a buffer is determined by the ratio of the salt to the acid present. The determination of the pH of a solution is one of the functions of a buffer. The other function is the protection of the system against pH changes. It does this by virtue of the high concentrations of salt (OAc^-) and weak acid (HOAc) used in the buffer (about 0.1 M, for example). If strong acids are added, they are neutralized by the acetate ion to form HOAc, an essentially undissociated acid,

$$H_3O^+ + OAc^- \longrightarrow HOAc$$

and if strong bases are added, they are neutralized by the acetic acid to form OAc^-:

$$OH^- + HOAc \longrightarrow H_2O + OAc^-$$

In either case, the strong acid or base is neutralized without an appreciable change in the ratio of OAc^-/HOAc, and consequently in the pH of the system.

Suppose we compare the effects of adding 10^{-3} moles of HCl (0.037 g) to 1 liter of a dilute acid solution (10^{-5} M HCl), and also to 1 liter of a solution buffered with 0.175 M NaOAc and 0.100 M HOAc, both of which solutions have an initial pH of 5.

Nonbuffered system (10^{-5} M solution of HCl):

Initial pH = 5, since $[H_3O^+] = 10^{-5}$

Final pH: $[H_3O^+] = 10^{-5} + 10^{-3} = 0.00101 = 1.01 \times 10^{-3}$, and pH = 2.996

Total pH change = 5 − 2.996 = 2.004 pH units

Buffered system (0.175 M NaOAc + 0.100 M HOAc):

$$\text{Initial pH} = pK + \log\frac{[OAc^-]}{[HOAc]}$$

$$= -\log 1.75 \times 10^{-5} + \log\frac{0.175}{0.100}$$

$$= 4.757 + 0.243 = 5.000$$

$$\text{Final pH} = pK + \log\frac{(0.175 - 0.001)}{(0.100 + 0.001)} \qquad [OAc^- + H_3O^+Cl^- \longrightarrow$$

$$HOAc + H_2O + Cl^-]$$

$$= 4.757 + \log\frac{0.174}{0.101}$$

$$= 4.757 + \log 1.72$$

$$= 4.757 + 0.236 = 4.993$$

Total pH change = 5 − 4.993 = 0.007 pH units

The pH change in the unbuffered system is large (equivalent to a hundredfold increase in the hydronium ion concentration), whereas the pH change in the buffered system is negligible.

A large number of different acid–salt pairs are used as buffers in chemical and biological systems; examples are $H_3PO_4 + NaH_2PO_4$, $NaH_2PO_4 + Na_2HPO_4$, $NaHCO_3 + Na_2CO_3$, and $NH_3 + NH_4Cl$.

Energy is the capacity to do work, with *work* ultimately being defined in terms of a force times a displacement. The work done in sliding a block along a surface, for example, is equal to the product of the force required to move the block and the distance it is moved.

Many kinds of energy are familiar to the reader: mechanical energy, electrical energy, heat energy, chemical energy, and so on. The different kinds of energy are, with few exceptions, readily interconverted. This conversion is relatively straightforward in mechanical systems (systems involving the movement of matter). Mechanical energy is conveniently subdivided into two types: potential energy (stored energy, or energy of position) and kinetic energy (energy of motion). A barrel of water on a mountain top has a certain amount of potential energy, which is proportional to the height of the mountain (and, of course, to the quantity of water in the barrel). If the water is poured over the edge of a cliff, the potential energy drops as the water falls, but the kinetic energy increases. When the water strikes the floor of the canyon, at which point the potential energy $= 0$, an amount of heat energy is released equal to the kinetic energy of the water just before it reached the canyon floor, and this amount, in turn, is equal to the potential energy the water had on the mountain top (we neglect the effect of air friction and otherwise assume ideal conditions). If the water is allowed to strike a paddle wheel, on the other hand, a part of the energy may be obtained as work, although the rest is again converted into heat. Energy, therefore, may be converted from one form to another but in these transformations, the total amount of energy remains constant. This is essentially a statement of the first law of thermodynamics, which is also quoted in the following form: "Energy may be neither created nor destroyed."

However, some specific energy conversions may be inefficient; the conversion of electrical energy into light energy in a light bulb is only a few percent efficient. However, the rest of the electrical energy is converted at the same time into heat. The point is that the *sum* of the light energy and heat energy (output) is equal to the electrical energy input.

We are particularly interested here in the energy that can be obtained from chemical compounds (that is, in chemical energy) since muscular movement, nerve impulses, growth, and all other biological processes (with the exception of photosynthesis) are "run" by the chemical energy of the food ingested by the organism. Chemical energy can be defined as the energy a molecule has by virtue of the kinds of atoms it contains and the manner in which they are linked together; it may be considered as a type of potential energy. Explosions and flames are dramatic examples of reactions in which chemical energy is released. The heat and light that are characteristic of these reactions come from the breakdown or rearrangement of the chemicals

CHEMICAL REACTIONS

involved; that is, the energy released was present originally in the starting molecules as chemical energy.

Silver azide is an explosive compound that breaks down into silver atoms and nitrogen molecules; in the process it liberates considerable energy in the form of heat and light, thus:

$$2 \; Ag \!-\! \overset{..}{N} \!\!=\!\! N \!\!=\!\! \overset{..}{N} : \longrightarrow 2 \; Ag + 3 : N \!\!\equiv\!\! N : + \text{heat} + \text{light}$$
Silver azide

The three nitrogen molecules and the two silver atoms contain much less chemical energy than did the two silver azide molecules because of the different ways in which the atoms are bonded, and in a general sense, the products are more stable than are the reactants. The energy difference between reactants and products is the energy liberated during the explosion.

If the explosion of silver azide is carried out in the piston chambers of an engine, work as well as heat can be had from the explosion. It has been found that the sum of the heat (q) and the work (w) from a chemical reaction is a constant. When the experiment is carried out at a constant pressure and "work" (w') includes all of the work done except that of expansion of gaseous products against the atmosphere, the constant is given the symbol ΔH ("delta H") and is called the *heat of reaction*. Thus, $\Delta H = q + w'$. The equation can be worded as "The loss of chemical energy in a reaction is equal to the sum of the heat lost and the work done." The value of ΔH depends on the concentrations of the reactants and products, and also on their *state*, that is, on whether they are liquids, solids, or gases. It is customary to standardize these variables and we will be concerned with compounds in the form normally found at 1 atmosphere (atm) pressure and 25°C, or with solutions in water at unit concentration (effectively 1 M); a superscript zero is attached to the symbols used to indicate these standard conditions (for example, ΔH^0).

The value of ΔH^0 can be easily determined when the reaction is carried out at a constant pressure and when no work other than gas expansion work is done ($w' = 0$). Under these conditions, $\Delta H^0 = q$. The heat (q) can be easily measured in a calorimeter. Thus, when 1 mole of methane (CH_4) is burned in air, exactly 213,000 calories° (cal) of heat energy is released.

$$1 \; CH_4 + 2 \; O_2 \longrightarrow 1 \; CO_2 + 2 \; H_2O$$
$$\Delta H^0 = -213,000 \; \text{cal}$$

The negative sign here for ΔH^0 indicates that the heat is *released* by the reaction. A few reactions have a positive ΔH^0 and during the course of these reactions, heat is absorbed from the surroundings. These two types of reactions are called *exothermic* and *endothermic*, respectively.

° A *calorie* (cal) is that quantity of heat required to raise the temperature of one gram of water from 14.5°C to 15.5°C.

The heat energy of a reaction can be used to perform work in a mechanical device. For example, heat from the burning of methane (natural gas) can generate steam to run a steam engine, which in turn can perform mechanical work. We shall see later that chemical reactions occurring in living cells can also perform work, as in the stretching of a muscle fiber. No machine is 100 percent efficient, which means that not all of the energy input of a machine can be converted into work; a large fraction is lost as heat, light, and other forms of energy. Also, in living systems (for example, in bacteria) only a part of the chemical energy of the food intake can be utilized for growth, movement, and so on; the rest is lost, usually as heat. Of paramount importance, therefore, is the question: "What is the maximum amount of work that can be obtained from a chemical reaction?" This quantity can be determined easily if we specify certain conditions—namely, that the reaction be carried out at a constant temperature and pressure and in a reversible manner. The latter term means that we should be able to shift the reaction from the forward to the backward direction by a very slight change in one of the variables. Under these conditions, $\Delta H^0 = q_{rev} + w'_{rev}$. By convention, w'_{rev} is given the symbol ΔG^0 (called the *free energy change*), which represents the maximum work that the reaction or system can do. The heat term, q_{rev}, is given the symbol $T \Delta S^0$, where T is the absolute temperature (°Kelvin = °Centigrade + 273) and ΔS^0 is a thermodynamic quantity called the *entropy change*. Our equation now reads $\Delta H^0 = T \Delta S^0 + \Delta G^0$, which can be rearranged to determine ΔG^0, the work term: $\Delta G^0 = \Delta H^0 - T \Delta S^0$. That is, if ΔH^0 and ΔS^0 are known for a reaction, ΔG^0, the maximum work available—or, in a sense, the usable energy of the reaction—can be calculated.

A feeling for the meaning of entropy can be obtained by examining a simple, reversible physical process, the melting of ice: $H_2O_{ice} \rightleftharpoons H_2O_{liq}$. It has been shown experimentally that the melting of 1 mole of ice at 0°C and 1 atm pressure requires the input of 1,447 cal of heat energy ($= q_{rev}$). Since no work is done in this process, $\Delta G^0 = 0$, and our equation $\Delta G^0 = \Delta H^0 - T \Delta S^0$ reduces to $T \Delta S^0 = \Delta H^0$. Since, too, $\Delta H^0 = q_{rev} = 1,447$ cal, the entropy change can be readily calculated:

$T \Delta S^0 = 1,447$ cal and $\Delta S^0 = 1,447/T = 1,447/0° + 273$
$= 1,447/273° = 5.3$ cal/mole-degree

That is,

$\Delta S^0 = S^0_{liq} - S^0_{ice} = +5.3$ cal/mole-degree

The entropy of a system tells us how disordered that system is. The water molecules in an ice crystal are arranged in a very regular way, just as the sodium and chloride ions are arranged regularly in a crystal of sodium chloride (Figure 1.7). In liquid water, on the other hand, the arrangement of water molecules is more haphazard and the molecules are essentially free to move about. Ice crystals and other crystals with ordered structures have low entropies, whereas the corresponding liquids with more random structures have high entropies. In general, systems tend to move from a more ordered structure to a less ordered one, and in this process the entropy increases. The entropy can also be considered a measure of the nonuseful, or nonavailable, energy in a system. The general equation $\Delta G^0 = \Delta H^0 - T \Delta S^0$ (on rearrangement, $\Delta H^0 = \Delta G^0 + T \Delta S^0$) can be worded as "The chemical energy of a reaction equals the work, or useful energy of that reaction, plus nonavailable energy."

Two general methods exist for calculating the maximum amount of work obtainable from a reaction. The first requires measurement of ΔS^0 and ΔH^0. For example, these quantities have been determined for the decomposition of nitrous oxide (N_2O): $N_2O \longrightarrow N_2 + \frac{1}{2} O_2$; $\Delta H^0 = -19,500$ cal; $\Delta S^0 = +18$ cal per mole degree; $T = 25°C$ or $298°$ Kelvin. From our equation for the free-energy change, $\Delta G^0 = \Delta H^0 - T \Delta S^0 = -19,500 - 298 \; (18) = -24,864$ cal. If a battery were constructed based on this reaction, the maximum electrical work that we could obtain from it would be equivalent to 24,864 cal. The oxidation of glucose in living cells provides another example:

Glucose $+ 6 O_2 \rightarrow 6 CO_2 + 6 H_2O$
$\Delta H^0 = -673,000$ cal/mole
$\Delta S^0 = +43.6$ cal/mole-degree
Calculate ΔG^0. (Answer: $-686,000$ cal/mole.)

Thus, the digestion of glucose makes available a large amount of energy.

A second, more direct method for obtaining ΔG^0 can be used if the reaction reaches equilibrium, since a simple relationship exists between ΔG^0 and the equilibrium constant K: $\Delta G^0 = -2.3 \; RT \log K$. The "R" in the equation is a constant with the value 1.99 cal/mole-degree. As an example, consider the rearrangement of glucose-1-phosphate into glucose-6-phosphate, one of the steps in the biological oxidation of glucose mentioned above (structures for these compounds are shown in Chapter 6).

Glucose-1-phosphate $\rightleftharpoons$ glucose-6-phosphate
(0.001 M)   (0.019 M)

$$K = \frac{[\text{glucose-6-phosphate}]}{[\text{glucose-1-phosphate}]} = \frac{0.019}{0.001} = 19$$

The values under the equation are the concentrations of the two compounds at equilibrium. The very same equilibrium concentrations would be obtained whether we started with pure 0.020 M glucose-1-phosphate and waited until equilibrium was established, or whether we started with pure 0.020 M glucose-6-phosphate and again waited for the mixture to reach equilibrium. The

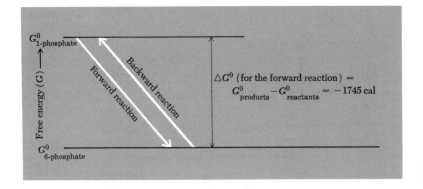

Figure 2.2 *Free-energy change* in the iso-merization of glucose-1-phosphate.

equilibrium constant for this system is 19, and substitution of this value of K into our free-energy equation leads to a ΔG^0 of $-1,745$ cal; by definition, this value means that when 1 mole of glucose-1-phosphate is converted into 1 mole of glucose-6-phosphate under standard conditions, the maximum amount of work that can be done is equal to 1,745 cal. The changes in free energy occurring in this reaction are illustrated in Figure 2.2. Glucose-1-phosphate has a higher free energy than has glucose-6-phosphate, and the conversion of 1 mole of it into 1 mole of glucose-6-phosphate results in a decrease of 1,745 cal of energy. It is obvious that if we wished to convert 1 mole of glucose-6-phosphate into glucose-1-phosphate (an "uphill" process), it would be necessary to supply 1,745 cal of free energy from an outside source.

The relationship of the equilibrium constant for a reaction and the free-energy change can be outlined in the following way. Equilibria in which most of the reactant is converted into product (often indicated by the relative lengths of the arrows in the equilibrium, for example, A $\rightleftharpoons$ B) have large equilibrium constants and negative values of ΔG. On the other hand, equilibria in which very little of the reactant is converted into product (B $\rightleftharpoons$ A) have small equilibrium constants and positive values of ΔG.

The sign of ΔG thus tells us whether the reaction will proceed to an appreciable extent as written. However, although a negative ΔG means that the reaction will proceed as written, it tells us nothing about how fast the reaction will be.

Some reactions require a fraction of a second to reach equilibrium, others require centuries. Additional information is necessary before the *rate* at which equilibrium is reached can be discussed.

A few chemical reactions are very fast. The combination of iodine atoms in solution to form molecular iodine is an example ($:\ddot{\underline{I}}\cdot\ +\ \cdot\ddot{\underline{I}}: \ \longrightarrow\ :\ddot{\underline{I}}:\ddot{\underline{I}}:$). In this case, virtually every collision of the iodine atoms leads to an iodine molecule.

Most reactions of simple ions are also very fast:

$$I^- + Ag^+ \longrightarrow AgI \quad \text{(precipitate)}$$

Most of the reactions of organic compounds of biological interest are comparatively slow, on the other hand, and the time needed for the reactions to proceed to 50 percent completion ranges from seconds to years, depending on the particular chemicals involved. Why should these reactions be so slow? One of the reasons is because not every collision of large atoms or molecules can lead to a chemical reaction. If, for example, we are concerned with the ionization of a methyl alcohol molecule:

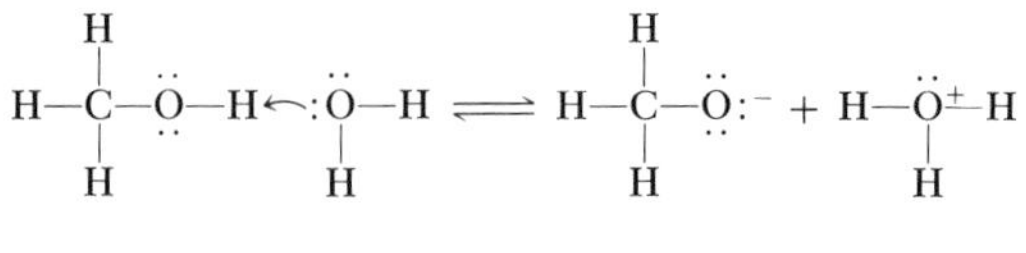

the collision of a water molecule on the wrong side of the molecule obviously cannot lead to the CH_3O^- ion since only the proton directly attached to the oxygen atom can be transferred to the water molecule:

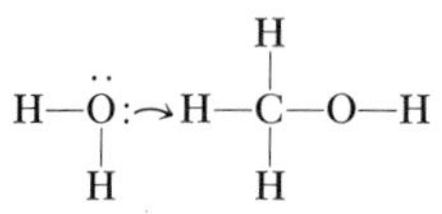

A second reason why reactions are slow is because molecules must gain some additional kinetic energy from collisions with neighboring molecules in order to overcome the repulsion of the electrons on the two reactants. Only very high-speed (high-energy) collisions between reactants allow chemical changes to take place. These effects lead to a free-energy barrier between reactants and products. This point is illustrated in Figure 2.3. The free-energy barriers (G of activation) determine the speed of reactions; the higher the barrier, the slower the reaction. Path c represents a reaction mode with zero activation energy (no barrier), and a reaction following this path would be extremely fast—so fast in fact that the reaction would be over in approximately one-millionth of a millionth of a second for 1 M concentrations of the reactants. The conversion of glucose-1-phosphate into glucose-6-phosphate is a slow reaction, however, and it follows another path (path a), which contains a reasonably large energy barrier. This barrier can be decreased and the rate of the reaction increased through the use of a catalyst, as we shall see in the next section. An incease in the rate of a reaction may also be brought about by increasing the temperature or the concentrations of the reactants; the references listed in the bibliography should be consulted for further information about the latter variables.

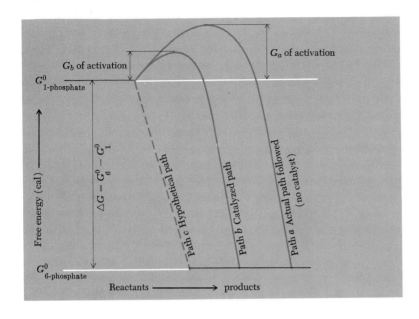

Figure 2.3   *Three possible paths with different rates for the conversion of glucose-1-phosphate into glucose-6-phosphate.*

CATALYSTS   Certain substances called *catalysts* are able to lower the energy of activation of a reaction. In Figure 2.3 this is illustrated by the effect of a catalyst in lowering the barrier from the value in path *a* to that in path *b*. The catalyzed reaction proceeds much faster than did the previous reaction since more molecules are able to pass the lower barrier. A complex protein catalyst called an enzyme (p. 173) was used to catalyze the glucose-1-phosphate reaction. Many other simple compounds, such as acids, bases, platinum metal, and so on, can act as catalysts in chemical reactions. The rate at which carbonic acid is formed from CO_2 and H_2O,

$$CO_2 + H_2O \xrightarrow{H_3O^+} H_2CO_3$$

is increased markedly by adding acids, for example. Catalysts work in various ways but in general it appears that they form a weak complex with the reactants and that after the energy barrier is passed, the catalyst is regenerated unchanged. That is, a catalyst is not consumed during a reaction; for this reason, only very small amounts of catalysts usually are required. More detailed examples of catalysis are given in Chapter 3.

ENERGY UTILIZATION IN BIOLOGICAL SYSTEMS   Living organisms are extremely complex systems that are "run" by the chemical energy in the food ingested. The specific energy measure of interest here is the free energy of the food. A delicate balance exists between the free-energy input and the free-

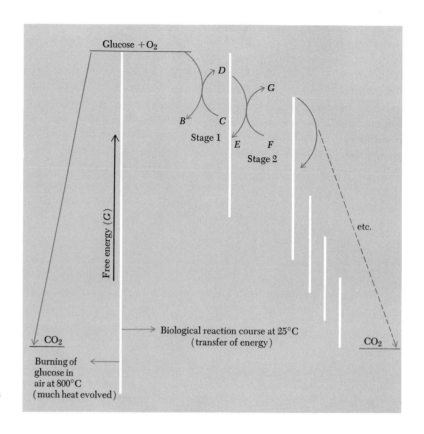

Figure 2.4 *Two paths for the oxidation of glucose to CO_2.*

Labels within figure:

Glucose $+O_2$

Free energy (G)

D

G

B C

Stage 1

E F

Stage 2

etc.

CO_2

Biological reaction course at 25°C
(transfer of energy)

CO_2

Burning of glucose in air at 800°C
(much heat evolved)

energy output (heat, motion, energy of excreted products, and so on), and if the energy output greatly exceeds the input for any appreciable period of time, the organism dies. For this reason, a living organism must make efficient use of the energy in its food.

Although a steam engine can convert a portion of any heat input into work, an organism or a cell has no way of putting heat generated by the digestion of food to work. Instead, the organism makes direct use of the free energy available in the food by transferring, at each of many discrete steps, the available free energy in the form of chemical-bond energy. A series of coupled reactions take place (Figure 2.4) in which, at each stage, a low-energy molecule from the surroundings is converted into a relatively high-energy molecule at the expense of the free energy of the oxidation.

For example, in stage 1 the glucose is partially oxidized to molecule B, which has lower energy, and species C picks up part of the energy, becoming converted into D, which has a high enough energy to drive stage 2, and so on.

It is necessary that the product D of stage 1 is a reactant in stage 2, that the product G is a reactant in stage 3, and so on. In this way, whole groups of atoms (for example, the phosphate group) can be transferred from one molecule to another down the chain and can act as a type of energy carrier. Not only is the energy transferred efficiently in the oxidation, but in addition, new high-energy species (D, G, and so on) are formed, and these can enter into other vital reactions elsewhere in the organism. The extra energy in D and G usually is localized in one or two chemical bonds, which often are bonds to inorganic phosphate groups. These high-energy phosphate bonds and the nature of the coupled energy transfers will be outlined further in Chapters 4 and 5.

3 ORGANIC CHEMISTRY: THE HYDROCARBONS

ORGANIC CHEMISTRY IS THE CHEMISTRY OF CARBON COMPOUNDS. THE NAME WAS assigned in the early part of the eighteenth century when a distinction was made between inorganic chemistry, which dealt with the mineral kingdom, and organic chemistry, which dealt with the plant and animal kingdoms. At that time, organic compounds had been isolated only from plant and animal sources and, indeed, the synthesis of organic compounds from carbon, hydrogen, and inorganic compounds was considered impossible. The synthesis of urea by Wohler in 1828 and of acetic acid by Kolbe in 1845 served to overthrow these ideas, and today we recognize that although carbon compounds are essential for life, there are no intrinsic differences, in a philosophical sense, between inorganic and organic compounds.

The number of carbon compounds known today exceeds the number of compounds prepared from the other 102 elements. This complexity of carbon chemistry is a result of three properties of the element: the high covalency of carbon (4), which permits a large number of groups to attach to it in a great variety of combinations; the great strength of the carbon-carbon bond, which permits the formation of chains of carbon atoms of unlimited length; and the formation of multiple bonds by carbon, which further increases the number of possible organic compounds. A few of the other elements have one or two of these characteristics, but none has all three.

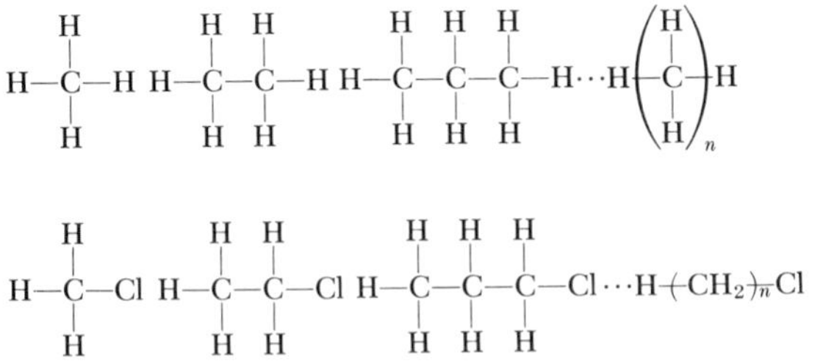

Figure 3.1 **A homologous series** of alkanes, and a homologous series of chlorine compounds.

The organization of organic chemistry was vastly simplified by the recognition that there are *homologous series* of compounds; in these series, the members differ only in the number of building blocks (such as CH_2 groups) per molecule (Figure 3.1). In general, the members of a homologous series have similar properties that are very often determined not by the carbon chain, which is rather inert, but by some small group (such as OH, Cl, NO_2, $C\equiv CH$, and so on) called a *functional group*, which is attached to the chain. In this volume we shall organize our discussion primarily by functional groups; the number of specific compounds mentioned will be kept to a minimum through the use of a few examples and through the extrapolation of the facts covered to other members of each homologous series.

This chapter on organic chemistry will deal with the hydrocarbons, defined as compounds containing only carbon and hydrogen, and the subject will be subdivided in the following way:

I.   Aliphatic hydrocarbons
    A.   Alkanes—compounds related to methane, CH_4
    B.   Alkenes—compounds related to ethylene, $CH_2{=}CH_2$
    C.   Alkynes—compounds related to acetylene, $CH\equiv CH$
II.   Aromatic hydrocarbons—compounds related to benzene C_6H_6

The alkanes (also called saturated hydrocarbons or paraffins) are those hydrocarbons with a maximum ratio of hydrogen to carbon (general formula C_nH_{2n+2}).

The simplest alkane, CH_4, is the principal constituent of natural gas. More complex hydrocarbons have been isolated from natural sources, such as beeswax and the plant waxes. Mixtures of hydrocarbons ranging in size from approximately $C_{15}H_{32}$ to $C_{35}H_{72}$ have been found on the surfaces of the leaves of many plants and also on fungal spores; their role here seems to involve the blocking of rapid water losses since the hydrocarbons are impervious to water. Similar mixtures of hydrocarbons have been isolated from geological strata of the Pre-

Table 3.1   *Straight-chain alkanes*

FORMULA	NAME	MELTING POINT, °C	BOILING POINT, °C
CH_4	*Methane*	−183	−161
C_2H_6	*Ethane*	−172	−89
C_3H_8	*Propane*	−188	−43
C_4H_{10}	*n-Butane*	−137	0
C_5H_{12}	*n-Pentane*	−130	36
C_6H_{14}	*n-Hexane*	−94	69
C_7H_{16}	*n-Heptane*	−91	98
C_8H_{18}	*n-Octane*	−57	126
C_9H_{20}	*n-Nonane*	−54	151
$C_{10}H_{22}$	*n-Decane*	−30	174
$C_{11}H_{24}$	*n-Undecane*	−26	196
$C_{12}H_{26}$	*n-Dodecane*	−10	216
$C_{13}H_{28}$	*n-Tridecane*	−6	232
$C_{20}H_{42}$	*n-Eicosane*	37	>232
$C_{21}H_{44}$	*n-Heneicosane*	40	>232
$C_{22}H_{46}$	*n-Docosane*	44	>232
$C_{30}H_{62}$	*n-Triacontane*	68	>232
$C_{40}H_{82}$	*n-Tetracontane*	81	>232

cambrian period, where, it is thought, they represent the remains of life as it existed over a billion years ago. In the intervening time, all of the other less stable constituents of cells, such as sugars, proteins, DNA, and so on, have decomposed, and only the hydrocarbons remain.

Although hydrocarbons are widely distributed in plants and animals, the amounts found are usually small. Most hydrocarbons are therefore obtained from petroleum, a relatively abundant material made up of a complex mixture of various hydrocarbons and certain other compounds containing oxygen, nitrogen, sulfur, and small amounts of the other elements.

The simplest homologous series of alkanes is the straight-chain series, which as the name implies contains members with a single chain of CH_2 groups capped at each end by a hydrogen atom (hence the general formula C_nH_{2n+2}). A listing of some members of this series is given in Table 3.1. The lower members are assigned specific names (devised before the existence of homologous series was recognized) whereas the higher members are assigned systematic names based on the Greek and Latin numerical prefixes and the general suffix *-ane*, which stands for saturated hydrocarbons; the prefix *n-* (normal) indicates that the molecule has a straight unbranched chain.

THE SHAPES OF ORGANIC MOLECULES   The shapes of the alkanes may be deduced from the shape of the methane molecule. In Chapter 2, we saw that the four bonds in methane are directed to the corners of a regular tetrahedron;

this is represented in Figure 3.2. The higher hydrocarbons are made up of chains of these tetrahedral units.

Simpler methods are available for indicating the geometric configurations of the hydrocarbons; three of these methods are given in Figure 3.3. The drawings in the first column of Figure 3.3 are based on models of the methane molecule in which the diameters of the carbon and hydrogen atoms, and the bond lengths, are constructed to scale as accurately as possible. The figures in the second column represent "ball-and-stick" models, which illustrate more clearly the bond angles involved. The figures in the third column are "projection" formulas that are so drawn as to aid in visualizing the geometry of the molecules. In addition to these methods of representation, we very often for convenience use graphic formulas, such as

$$H-\overset{\displaystyle H}{\underset{\displaystyle H}{C}}-H$$

(which may lead the unwary to believe that molecules are planar!), and molecular formulas, such as CH_4, C_2H_6, and so on. In order to grasp more fully the

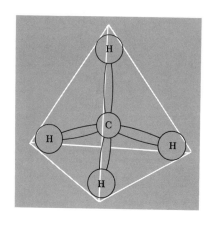

Figure 3.2 *A methane molecule inscribed in a regular tetrahedron.*

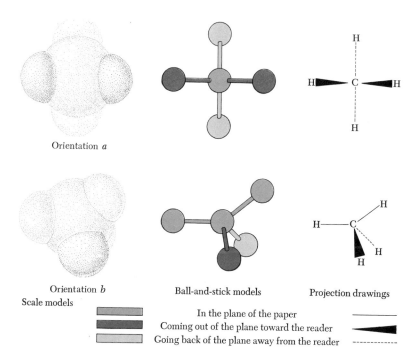

Figure 3.3 *Models and drawings of the methane molecule.*

Orientation *a*

Orientation *b*
Scale models

Ball-and-stick models

Projection drawings

In the plane of the paper
Coming out of the plane toward the reader
Going back of the plane away from the reader

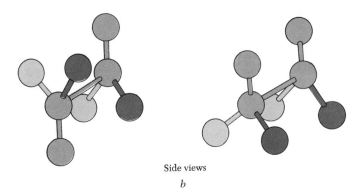

Staggered form   Eclipsed form

a

Figure 3.4   Conformations of ethane,
C_2H_6 *or*

$$H—\underset{\underset{H}{|}}{\overset{\overset{H}{|}}{C}}—\underset{\underset{H}{|}}{\overset{\overset{H}{|}}{C}}—H$$

Side views

b

structures represented by the projection or graphic formulas, models° of these compounds should be constructed where they first appear in the text.

A set of drawings representing the ethane molecule is given in Figure 3.4. Certain physical measurements indicate that at room temperature, essentially free rotation about single bonds (σ bonds) is possible. In this rotation about the single bond in ethane, two extreme structures are possible. The first set of drawings in Figure 3.4 represents an ethane molecule in which the hydrogens on adjacent carbon atoms are as far apart as possible (see end view); this is the *staggered* form of ethane. The second set represents an ethane molecule in which the hydrogens are as close as possible; this is the *eclipsed* form of ethane. Forms of a molecule that differ only by rotation about single bonds

° Models for the simpler compounds can be constructed of wire. Note that an inexpensive set of molecular models is available from the publisher: *Framework Molecular Models Student Set* (Box 903, Prentice-Hall, Inc., Englewood Cliffs, N.J. 07632).

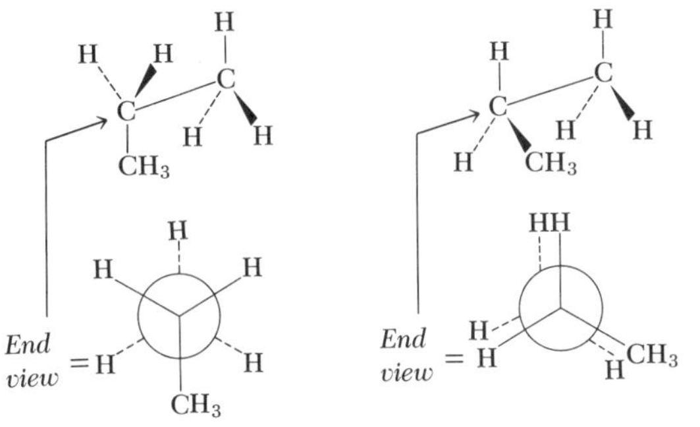

End view = H

Staggered form

End view = H

Eclipsed form

Figure 3.5 *Conformations of propane, C_3H_8 or*

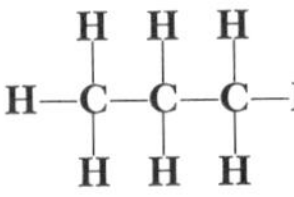

$$H-\overset{\overset{\displaystyle H}{|}}{\underset{\underset{\displaystyle H}{|}}{C}}-\overset{\overset{\displaystyle H}{|}}{\underset{\underset{\displaystyle H}{|}}{C}}-\overset{\overset{\displaystyle H}{|}}{\underset{\underset{\displaystyle H}{|}}{C}}-H$$

are called *conformations* of that molecule. The eclipsed conformations are the highest energy forms, whereas the staggered conformations are the lowest energy forms obtained by rotation about the C—C bond. As a result, the molecules exist almost entirely in staggered conformations at room temperature.

Similar representations of the conformations of propane and *n*-butane are given in Figure 3.5 and 3.6. If we extrapolate the carbon backbone of the most stable form of butane to the longer chain hydrocarbons, we find that the lowest energy form is the extended one in which the carbon atoms trace out, to a first approximation, the points on a saw (Figure 3.7).

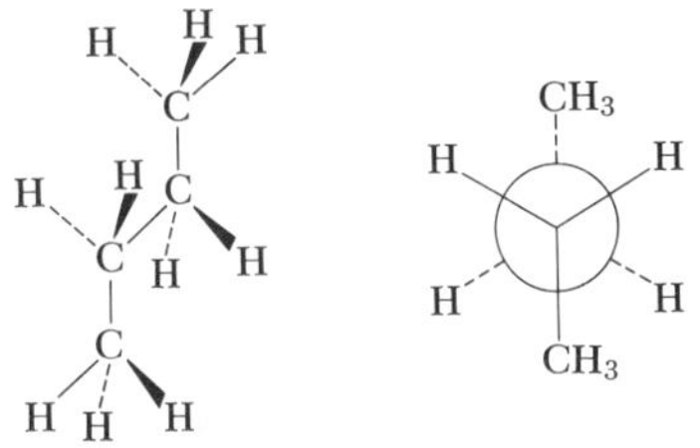

Figure 3.6 *The most stable conformation of n-butane, C_4H_{10} or*

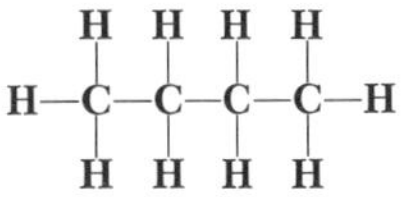

$$H-\overset{\overset{\displaystyle H}{|}}{\underset{\underset{\displaystyle H}{|}}{C}}-\overset{\overset{\displaystyle H}{|}}{\underset{\underset{\displaystyle H}{|}}{C}}-\overset{\overset{\displaystyle H}{|}}{\underset{\underset{\displaystyle H}{|}}{C}}-\overset{\overset{\displaystyle H}{|}}{\underset{\underset{\displaystyle H}{|}}{C}}-H$$

Figure 3.7 *(a) The most stable conformation of the carbon chain in the high-molecular-weight alkanes. (b) A molecular model of this carbon chain.*

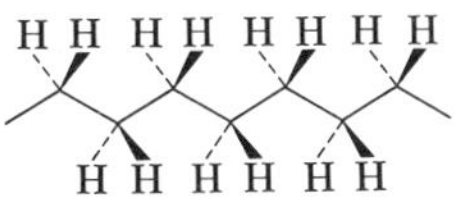

a

b

61

Figure 3.8   **Hypothetical** path for the preparation of a branched-chain hydrocarbon.

Propane → H· + Propyl radical

Methane → H· + Methyl radical

H_2 + 2-Methylpropane

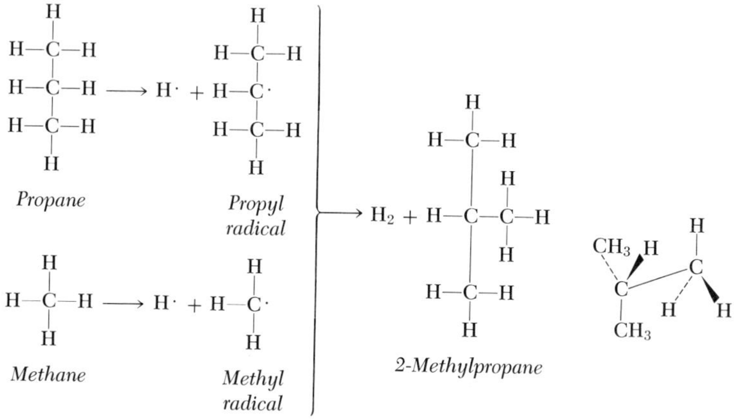

Figure 3.9   **Branched-chain hydrocarbons.**

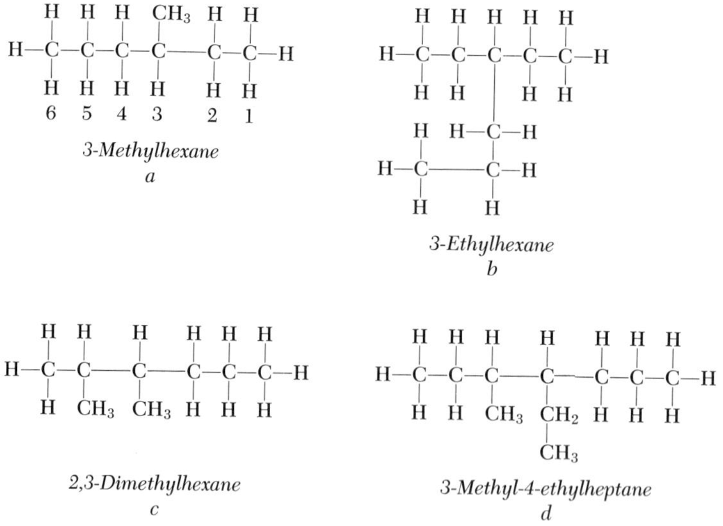

3-Methylhexane
a

3-Ethylhexane
b

2,3-Dimethylhexane
c

3-Methyl-4-ethylheptane
d

Figure 3.10   **Cyclic hydrocarbons.**

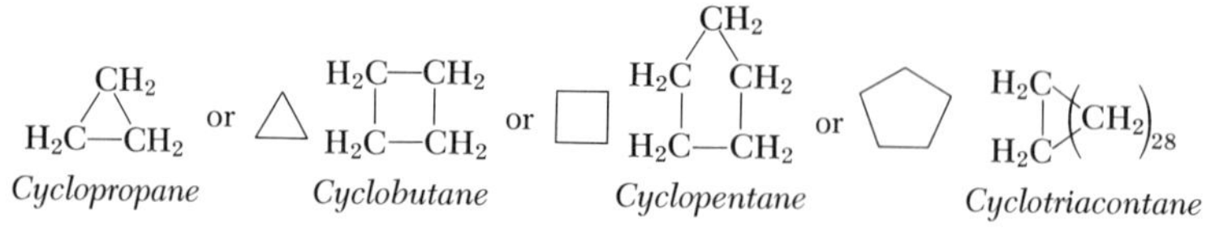

Cyclopropane   or   Cyclobutane   or   Cyclopentane   or   Cyclotriacontane

Figure 3.11   **Complex cyclic hydrocarbons.**

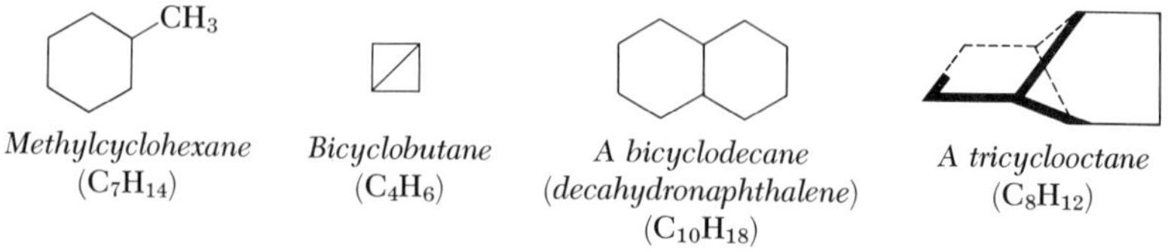

Methylcyclohexane   Bicyclobutane   A bicyclodecane   A tricyclooctane
(C_7H_{14})   (C_4H_6)   (decahydronaphthalene)   (C_8H_{12})
                                      ($C_{10}H_{18}$)

THE BRANCHED-CHAIN HYDROCARBONS   Many alkanes contain carbon branches attached to the long chain. The simplest branched-chain hydrocarbon is given in Figure 3.8, along with a hypothetical scheme for making the compound that illustrates how these compounds are named. The radical (or group) formed by the loss of a hydrogen atom from a hydrocarbon is named by dropping the alkane ending *-ane* and adding the ending *-yl*. Complex hydrocarbons are then named as radical derivatives of the longest chain of carbon atoms. For example, Figure 3.9*a* is named as a derivative of hexane, not of butane or pentane. The carbon atoms of the longest chain are then numbered from one end of the chain to the other, and the positions of the substituents are given by citing that number; for example, Figure 3.9*a* is 3-methylhexane. The carbons are numbered from that end of the chain that gives the lowest set of numbers; that is, Figure 3.9*a* is 3-methylhexane and not 4-methylhexane. If two substituents are present, each is given a number, as in Figures 3.9*c* and 3.9*d*. Note that there can be only one monomethyl propane (Figure 3.8), and only one monomethyl *n*-butane (namely, 2-methylbutane). There are no methyl-substituted methanes or ethanes since substitution leads merely to ethane and propane, respectively. The occurrence of alkyl branches is very common in the hydrocarbon portion of most naturally occurring compounds (examples which may be found in this volume are camphor, cholesterol, and vitamin A).

ISOMERS   Note that both *n*-butane and 2-methylpropane have the same molecular formula, C_4H_{10}. Different compounds with the same molecular formula are called *isomers*. In this particular case, the isomers differ only with respect to the position of the methyl group in the chain; in the butane molecule, the methyl group is attached to the end of a propane chain, whereas in methylpropane, it is attached to the central atom. Isomerism of this type is called *structural isomerism*. Similarly, *n*-hexane, 2-methylpentane, 3-methyl-pentane, 2,3-dimethylbutane, and 2,2-dimethylbutane are all isomers with the molecular formula C_6H_{14}; they are referred to collectively as isomers of hexane. The number of possible structural isomers increases rapidly with molecular size; there are two isomeric butanes, three pentanes, five hexanes (listed above), nine heptanes, 18 octanes, 35 nonanes, and a calculated 62,491,178,805,831 structural isomers with the general formula $C_{40}H_{82}$! It would be very instructive for the reader to write out structures for the isomeric heptanes and octanes.

CYCLIC ALKANES   A third class of alkanes is made up of the cyclic hydrocarbons; a number of examples are given in Figure 3.10. These compounds are named by appending to the prefix *cyclo-* the name of the corresponding straight-chain alkane. The formulas of the *mono*cyclic alkanes correspond to the general formula C_nH_{2n}. Many types of cyclic alkanes have been prepared, some quite complex; examples are given in Figure 3.11. Since the chemical properties of cyclic alkanes are very similar to those of linear alkanes, we shall not discuss them separately.

PREPARATION OF THE ALKANES   Many of the alkanes may be obtained by the fractional distillation of petroleum. Most of the complex hydrocarbons do not exist in nature, however, and they must be synthesized in the laboratory. One convenient method makes use of the reduction of corresponding halogen derivatives (which in turn are derived from the corresponding alcohols):

$$H_3C-\underset{\underset{X}{|}}{\overset{\overset{H}{|}}{C}}-CH_3 + 2\,(H\cdot) \longrightarrow CH_3CH_2CH_3 + HX$$

In the equation, $H\cdot$ represents a reducing agent and X is one of the halogens, (chlorine, bromine, or iodine). Common reducing agents are sodium metal in liquid ammonia, zinc in acetic acid, and lithium aluminum hydride ($LiAlH_4$). The latter reagent, which was introduced into organic chemistry procedures in 1947, is one of the most valuable of all reducing agents. Several other instances of its use are cited in Chapters 5 and 6.

The simplest alkane, methane, is commonly produced by anaerobic bacteria, that is, by bacteria that live in the absence of air. Suitable environments are deep earth sediments, the rumen of cattle, and bogs. In the latter environment, action of anaerobic bacteria on plant residues yields methane, but some phosphine (PH_3) is also formed concurrently. Phosphine ignites spontaneously in air and under these circumstances also sets the methane on fire. This appears to be the accepted explanation of the "will-o'-the-wisp," so often seen in medieval days, but so rarely seen today. The source of the carbon atom of methane in several strains of anaerobic bacteria is CO_2. Since oxygen is not available to these bacteria (indeed, it is toxic to them), some energy source other than the normal oxidation of foodstuffs by oxygen is necessary; the anaerobic production of methane ($CO_2 \rightarrow CH_4$) appears to be the oxidation-reduction reaction that provides the energy for life. As an example of the diversity of life, some bacteria are able to ingest and metabolize methane and other hydrocarbons; they have been found in soil collected near gasoline stations.

REACTIONS OF THE ALKANES   The alkanes are relatively stable compounds and the variety of reactions they undergo is limited; some of the more common reactions are given below:

$$2\,C_2H_6 + 7\,O_2 \longrightarrow 6\,H_2O + 4\,CO_2$$

$$CH_4 + Cl_2 \longrightarrow HCl + CH_3Cl$$
$$\textit{Chloromethane}$$

$$CH_4 + 4\,Cl_2 \longrightarrow 4\,HCl + CCl_4$$
$$\textit{Excess} \qquad \textit{Tetrachloromethane,}$$
$$\textit{or carbon tetrachloride}$$

$$C_2H_6 + Br_2 \longrightarrow HBr + CH_3CH_2Br$$
$$\textit{Bromoethane}$$

All hydrocarbons react with an excess of oxygen at high temperature to give carbon dioxide and water in a process called burning, or combustion. In addition to serving as a source of heat, this reaction is often used in chemical analysis since the empirical formula of a hydrocarbon can be calculated from the weights of carbon dioxide and water obtained on combustion (see Chapter 2).

FREE-RADICAL REACTIONS   A majority of the reactions of the alkanes are *free-radical reactions*; this means that species with an odd number of electrons are formed as intermediates. Reactions of this type (for example, the chlorination of methane) are usually started by the symmetrical cleavage of an electron-pair bond:

Step 1 $:\overset{..}{\underset{..}{Cl}}:\overset{..}{\underset{..}{Cl}}: + light \longrightarrow 2 :\overset{..}{\underset{..}{Cl}} \cdot$

The free radical formed (a chlorine atom in this case) then reacts with methane by "plucking off" a hydrogen atom to give HCl and a new radical:

Step 2a $:\overset{..}{\underset{..}{Cl}} \cdot + H-\overset{\overset{\displaystyle H}{|}}{\underset{\underset{\displaystyle H}{|}}{C}}-H \longrightarrow :\overset{..}{\underset{..}{Cl}}-H + \cdot\overset{\overset{\displaystyle H}{|}}{\underset{\underset{\displaystyle H}{|}}{C}}-H$

In step 2b, the methyl radical ($CH_3\cdot$) abstracts a chlorine atom from Cl_2 to give methyl chloride and a chlorine atom:

Step 2b $H-\overset{\overset{\displaystyle H}{|}}{\underset{\underset{\displaystyle H}{|}}{C}}\cdot + :\overset{..}{\underset{..}{Cl}}-\overset{..}{\underset{..}{Cl}}: \longrightarrow H-\overset{\overset{\displaystyle H}{|}}{\underset{\underset{\displaystyle H}{|}}{C}}-Cl + :\overset{..}{\underset{..}{Cl}}\cdot$

It is evident that each time a chloromethane molecule is formed (step 2b), a chlorine atom is also formed, and that this species can then enter into step 2a again, and so on. Thus, a long chain of reactions occurs, in which steps 2a and 2b alternate. In principle, we would need to add only one chlorine atom to an equimolar mixture of chlorine and methane to convert all of the reactants into chloromethane and hydrogen chloride (steps 2a and 2b). That is, the overall reaction (obtained by adding the equations of steps 2a and 2b) would be $Cl_2 + CH_4 \longrightarrow HCl + CH_3Cl$, even though several different steps are required in the reaction. In practice, more than one chlorine atom is required, though, because free radicals are annihilated by combination:

Step 3a $2 :\overset{..}{\underset{..}{Cl}}\cdot \longrightarrow Cl-Cl$

Step 3b $2 H-\overset{\overset{\displaystyle H}{|}}{\underset{\underset{\displaystyle H}{|}}{C}}\cdot \longrightarrow H-\overset{\overset{\displaystyle H}{|}}{\underset{\underset{\displaystyle H}{|}}{C}}-\overset{\overset{\displaystyle H}{|}}{\underset{\underset{\displaystyle H}{|}}{C}}-H$

ORGANIC CHEMISTRY: THE HYDROCARBONS

Nevertheless, one chlorine atom is usually sufficient to lead to the formation of several thousand chloromethane molecules, and only very small quantities of light are required to bring about the full reaction of chlorine with methane or with the other alkanes (often with explosive violence). Reactions of this type are called *radical-chain reactions*; step 1 is usually referred to as the *initiation step*, steps 2a and 2b as the *propagation steps*, and steps 3a and 3b as the *termination steps*.

If an excess of chlorine is used, the CH_3Cl is itself chlorinated, and by choosing the correct amount of chlorine, CH_2Cl_2, $CHCl_3$ (chloroform), and CCl_4 can be formed. The free radicals involved in the chlorinations are unstable and they have very short lifetimes. Bulky, resonance stabilized radicals are relatively stable, however, and a few are even isolable at room temperature. Free radicals are also formed in certain biological reactions, especially in oxidations catalyzed by iron and other metal-organic complexes.

THE ALKENES

The alkenes (often called olefins or unsaturated hydrocarbons) are hydrocarbons containing one or more double bonds. They can be pictured as arising from alkanes by the loss of two hydrogen atoms from adjacent carbons, The two odd electrons on the carbons then form a two-electron bond.

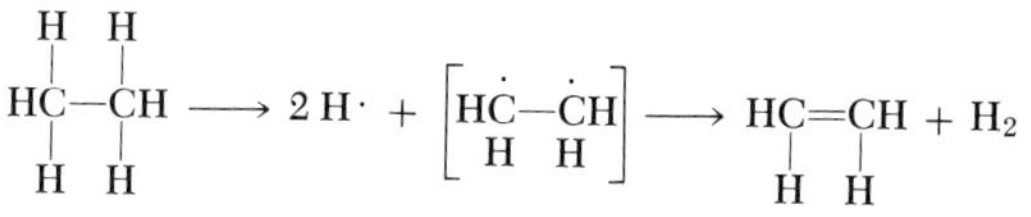

$$
\begin{array}{ccc}
\overset{H}{\underset{H}{\overset{|}{HC}}}-\overset{H}{\underset{H}{\overset{|}{CH}}} & \longrightarrow 2\,H\cdot + \left[\overset{\cdot}{\underset{H}{\overset{|}{HC}}}-\overset{\cdot}{\underset{H}{\overset{|}{CH}}}\right] & \longrightarrow \underset{H}{\overset{|}{HC}}=\underset{H}{\overset{|}{CH}} + H_2
\end{array}
$$

Ethane   *Ethene*

Typical alkenes are given in Figure 3.12. The alkenes are named by the substitution of the suffix *-ene* for the alkane suffix *-ane* (the simplest alkene is

Figure 3.12   **Representative alkenes.**

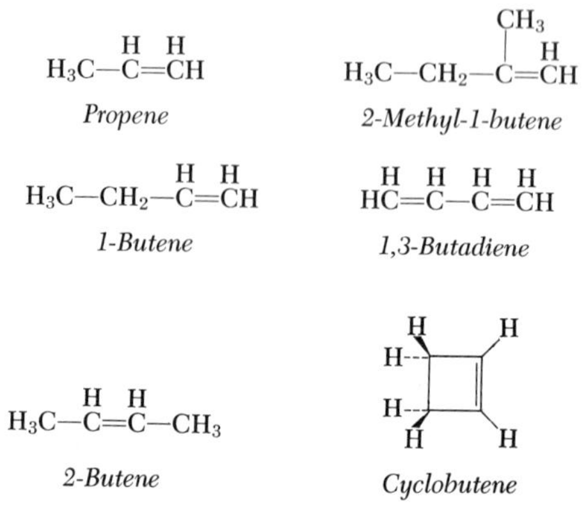

$$H_3C-\overset{H}{\overset{|}{C}}=\overset{H}{\overset{|}{CH}}$$
Propene

$$H_3C-CH_2-\overset{H}{\overset{|}{C}}=\overset{H}{\overset{|}{CH}}$$
1-Butene

$$H_3C-\overset{H}{\overset{|}{C}}=\overset{H}{\overset{|}{C}}-CH_3$$
2-Butene

$$H_3C-CH_2-\overset{CH_3}{\overset{|}{C}}=\overset{H}{\overset{|}{CH}}$$
2-Methyl-1-butene

$$\overset{H\ \ H\ \ H\ \ H}{HC}=\overset{|\ \ |\ \ |\ \ |}{C}-C=CH$$
1,3-Butadiene

Cyclobutene

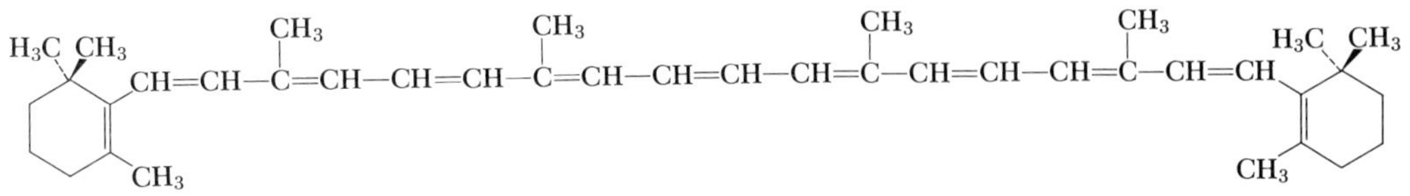

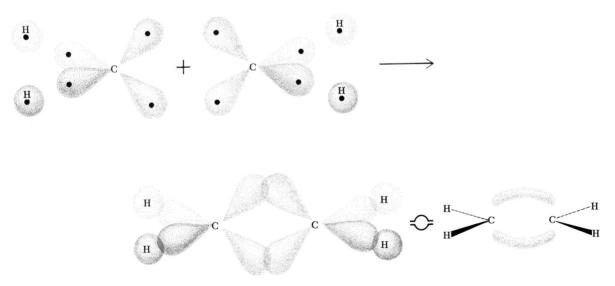

Figure 3.13   *β-Carotene.*

usually referred to as *ethylene*, however). In complex alkenes, in addition, a number is assigned to the first carbon atom of each double bond to indicate positions along the hydrocarbon chain (Figure 3.12).

The alkenes are quite common in nature; for example, the compounds responsible for the colors of tomatoes, carrots, boiled lobsters, and autumn leaves form a group of related polyolefins called *carotenes*. The carotenes are involved in photosynthesis, and they are intermediates in the biosynthesis of vitamin A and in other cellular processes. The structure of the most common carotene is given in Figure 3.13.

THE STRUCTURE OF THE ALKENES   The alkenes are far more reactive than are the alkanes, a fact that can be explained in terms of the structure of the double bond. We pointed out in Chapter 1 that the carbon atoms in methane are sp^3 hybridized. According to one group of theoretical chemists, the double bond is formed by the overlap of two sp^3 orbitals from each of two carbon atoms (Figure 3.14). The bonds joining the carbon atoms in this way are called *bent*, or *banana, bonds*.

Figure 3.14   **The double bond in ethylene** formed from two sp^3 hybridized carbon atoms (the dots symbolize electrons).

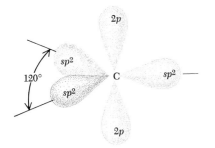

Figure 3.15   An sp² hybridized carbon atom.

According to a second group of theoretical chemists, the double bond is constructed of two sp^2 hybridized carbon atoms, thus:

$$\left[\begin{array}{l}\text{Electron} \\ \text{distribution} \\ \text{in an excited} \\ \text{carbon atom}\end{array}\quad \underset{2s\ 2p\ 2p\ 2p}{①\ ①\ ①\ ①}\right] \longrightarrow \left[\underset{\substack{\text{Three } sp^2 \\ \text{hybrid orbitals}}}{①\ ①\ ①}\quad \underset{2p}{①}\quad \begin{array}{l}\text{Electron} \\ \text{distribution} \\ \text{in an } sp^2 \\ \text{hybridized} \\ \text{carbon atom}\end{array}\right]$$

The three sp^2 orbitals in this atom are in one plane and are separated by 120°, whereas the $2p$ orbital is perpendicular to this plane (Figure 3.15). Two of the sp^2 bonds form bonds to the hydrogen atoms and the third forms a σ bond to the second carbon atom (Figure 3.16). The remaining $2p$ orbitals can then overlap to form a new molecular orbital called a π orbital (Figure 3.17).

Both physical pictures of the double bond lead to similar predictions, but the π-orbital picture of the double bond is the most widely accepted and we shall use it throughout the remainder of this volume. Since drawings such as Figure 3.17 are cumbersome, unless there is a necessity for emphasizing the π electrons, double bonds will be represented by two solid lines as in Figure 3.12.

The π electrons are more exposed than σ electrons are and this provides an explanation of why the alkenes are more reactive than the alkanes. The four hydrogen atoms in ethylene (and the carbon atoms attached to the double bond in the higher alkenes) lie in one plane and the π electrons are perpendicular to that plane. To rotate one carbon of ethylene by 90°, it would be

Figure 3.16   The formation of single bonds by two sp² carbon atoms.

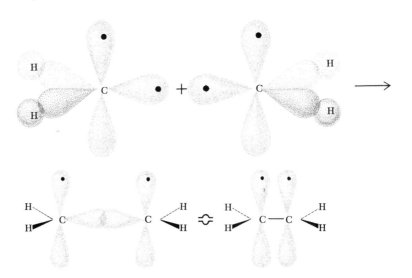

necessary to break the π bond to form two p orbitals in the new species (Figure 3.18). This would require a large amount of energy and consequently groups attached to a double bond are essentially locked in place. There is no rotation about a double bond at room temperature, in contrast to the essentially free rotation about single bonds. This analysis accounts for the existence of two stable isomeric 2-butanes (Figure 3.19a). In the *trans* isomer the two methyl groups are on opposite sides of the plane of the π electrons, whereas in the *cis* isomer they are on the same side of the plane. This type of isomerism is

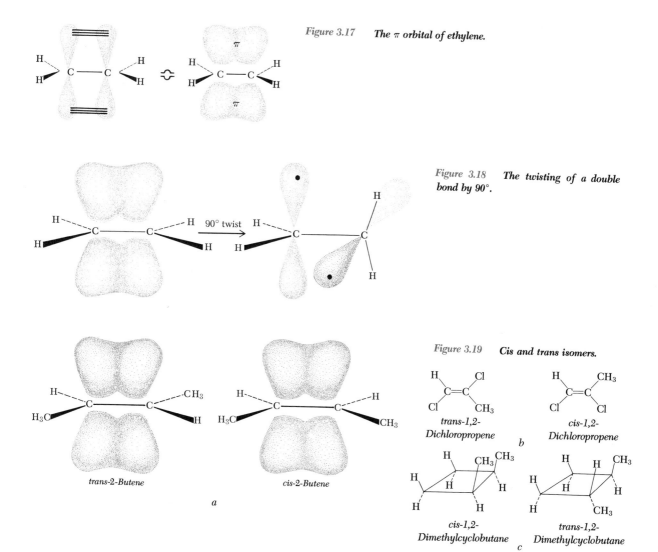

Figure 3.17   **The π orbital of ethylene.**

Figure 3.18   **The twisting of a double bond by 90°.**

90° twist

Figure 3.19   **Cis and trans isomers.**

trans-2-Butene

cis-2-Butene

a

H   Cl
\C=C/
Cl   CH₃

trans-1,2-
Dichloropropene

H   CH₃
\C=C/
Cl   Cl

cis-1,2-
Dichloropropene

b

cis-1,2-
Dimethylcyclobutane

trans-1,2-
Dimethylcyclobutane

c

called *geometric isomerism*; it occurs whenever four groups can be rigidly attached to different sides of a molecule. Two further examples are given in Figures 3.19*b* and 3.19*c*.

PREPARATION OF THE ALKENES   The alkenes are synthesized most often by removing the elements of a hydrohalic acid or water from the appropriately substituted alkane:

$$
\underset{\substack{| \\ H}}{\overset{\substack{H \\ |}}{H-C}}-\underset{\substack{| \\ H}}{\overset{\substack{Cl \\ |}}{C}}-\underset{\substack{| \\ H}}{\overset{\substack{H \\ |}}{C}}-H + KOH \longrightarrow H-\underset{\substack{| \\ H}}{\overset{\substack{H \\ |}}{C}}-C=\overset{\substack{H \\ |}}{C}-H + KCl + H_2O
$$

The latter reaction has been studied extensively and its mechanism has been well established. By *reaction mechanism* we mean a detailed account of a reaction in terms of the sequence and manner in which the bonds are made and broken. We have already examined one reaction mechanism, the chlorination of the alkanes (a free-radical chain reaction). The two reactions listed above in which alkanes are formed are called *elimination* reactions. They are also *ionic* reactions; that is, reactions in which ions are formed as intermediates. The mechanism of the acid-catalyzed elimination reaction is given in Figure 3.20.

In our representations of reaction mechanisms (Figure 3.20 and others), curved arrows indicate the direction of movement of electrons and also of molecules bearing unshared pairs of electrons. For example, in step 1 of Figure 3.20, the alcohol molecule collides with the hydronium ion in such an orientation that the unshared pair of electrons on the oxygen attacks a proton of the hydronium ion. The second arrow implies that the pair of electrons forming the O—H bond of the hydronium ion remains on the oxygen to yield a simple water molecule. The net effect of this reaction is the transfer of a proton to the alcohol molecule to yield a positively charged species very similar structurally to the hydronium ion. In the second step, the C—O bond is broken. This ionization reaction is made possible by protonation. The hydroxy compound itself does not ionize since this process would involve the separation of a negative ion (OH⁻) from a positive ion and this charge separation would require too much energy. In step 2, ionization yields the neutral species, H_2O, and a positively charged carbon ion, called a *carbonium ion*, containing only six electrons in its valence shell. Carbonium ions are highly reactive species that tend to gain the stable octet of electrons (neon electron structure) by

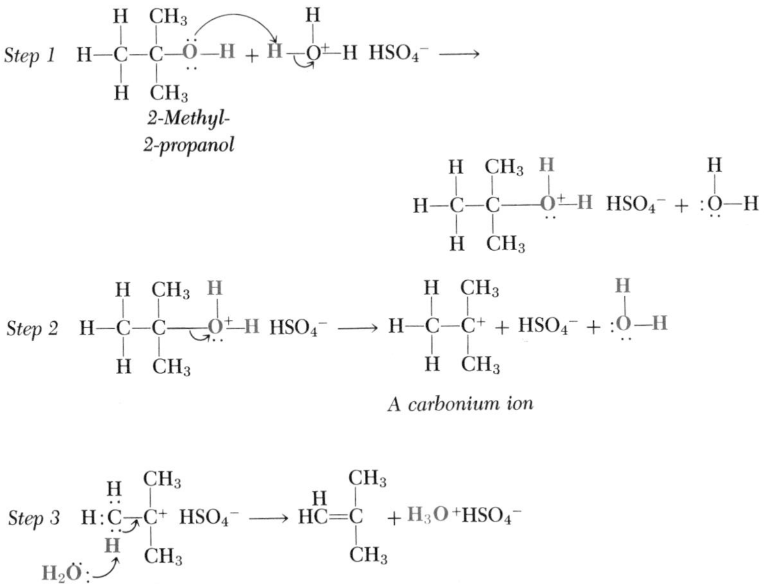

Step 1

$$\text{H—C—C—}\overset{..}{\text{O}}\text{—H} + \text{H—}\overset{..}{\underset{..}{\text{O}}}{}^{\pm}\text{—H}\ HSO_4^- \longrightarrow$$

2-Methyl-
2-propanol

$$\text{H—C—C——}\text{O}{}^{\pm}\text{—H}\ HSO_4^- + :\overset{..}{\text{O}}\text{—H}$$

Step 2

$$\text{H—C—C——}\text{O}{}^{\pm}\text{—H}\ HSO_4^- \longrightarrow \text{H—C—C}^+ + HSO_4^- + :\overset{..}{\text{O}}\text{—H}$$

A carbonium ion

Step 3

$$\text{H:C—C}^+\ HSO_4^- \longrightarrow \text{HC}=\text{C} + H_3O^+HSO_4^-$$

$$\text{H}_2\overset{..}{\text{O}}:$$

Figure 3.20 **The mechanism of the elimination of water** from 2-methyl-2-propanol.

"pulling in" electrons from an adjacent carbon-hydrogen bond to give an olefin (step 3, Figure 3.20), or by reacting with a negative ion present in solution. For example, if the *dehydration* (elimination of water) is carried out in the presence of chloride ion, 2-methyl-2-chloropropane is formed in addition to the olefin.

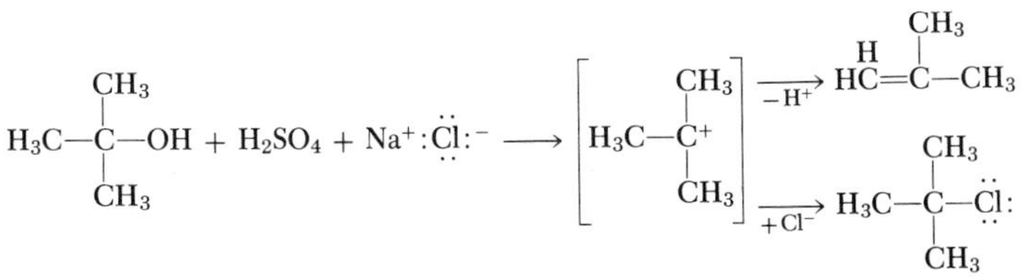

$$\text{H}_3\text{C—C—OH} + H_2SO_4 + Na^+:\overset{..}{\underset{..}{\text{Cl}}}:^- \longrightarrow \left[\text{H}_3\text{C—C}^+\right] \overset{-H^+}{\longrightarrow} \text{HC}=\text{C—CH}_3$$

$$\underset{+Cl^-}{\longrightarrow} \text{H}_3\text{C—C—}\overset{..}{\underset{..}{\text{Cl}}}:$$

Note that step 3 in Figure 3.20 regenerates a hydronium ion; thus the sulfuric acid (actually the hydronium ion) is acting as a true catalyst. It functions by decreasing the activation energy (energy barrier) for the cleavage of the C—O bond (step 2). This decrease, in turn, follows from the protonation of the alcohol, which allows the loss of a neutral water molecule to give the carbonium ion.

In contrast to this multistep reaction, the elimination of hydrohalic acids with bases is a concerted, or one-step, reaction; that is, a reaction in which no intermediates are formed. During the reaction, the C—H and C—Cl bonds are being stretched at approximately the same rate.

$$\text{H}\ :\!\overset{..}{\underset{..}{\text{Cl}}}\!:\ \text{H}$$
$$\text{H}-\overset{\displaystyle\text{H}}{\underset{\displaystyle\text{H}}{\text{C}}}-\overset{\displaystyle\text{H}}{\underset{\displaystyle\text{H}}{\text{C}}}-\overset{\displaystyle\text{H}}{\underset{\displaystyle\text{H}}{\text{C}}}-\text{H} + \ :\!\overset{..}{\underset{..}{\text{O}}}\!-\text{H} \longrightarrow \text{H}-\overset{\displaystyle\text{H}}{\underset{\displaystyle\text{H}}{\text{C}}}-\overset{\displaystyle\text{H}}{\text{C}}=\overset{\displaystyle\text{H}}{\text{C}}-\text{H} + \text{H}_2\text{O} + \ :\!\overset{..}{\underset{..}{\text{Cl}}}^-$$
$$\qquad\qquad\qquad\qquad\qquad\qquad\text{K}^+ \qquad\qquad\qquad\qquad\qquad\qquad\qquad\text{K}^+$$

The hydroxide ion attacks the proton in this reaction, and in light of our definition of acids and bases, we would say that it is acting here as a typical base.

The study of reaction mechanisms has shown that the very large number of organic reactions known today can be organized and correlated by a relatively small number of mechanisms. A knowledge of one reaction then permits the extrapolation of the information to other reactions of this type, and often to totally new reactions. Therefore, the mechanisms of important reactions will be outlined in each of the following sections, and a summary of the basic mechanisms will be given in Chapter 4.

Dehydration reactions are quite common in biological systems, where they are catalyzed by enzymes called hydrolyases. The mechanism of the biological dehydration is not known in detail, but it seems certain that it must bear some resemblance to the chemical dehydration outlined above insofar as the cleavage of the C—O bond is concerned. The total synthesis of alkenes such as carotene (Figure 3.13) in biological systems involves a very large number of steps that cannot be outlined here. It is interesting that microorganisms can synthesize β-carotene, using acetic acid as the sole carbon source. Through the use of acetic acid labeled in the methyl group with the radioactive isotope $_6\text{C}^{14}$, it has been possible to determine which atoms of carotene come from the methyl group and which from the carboxyl group (it has been found that the four methyl groups of the long chain stem from the methyl group of the acetic acid).

REACTIONS OF THE ALKENES   The chemical reactions of the alkenes are characterized by addition to the double bond; that is, by reactions in which the double bond is lost and derivatives of the alkanes are formed. Several examples follow; if a catalyst is required, it is indicated above the arrow in the equation.

$$\text{H}_3\text{C}-\overset{\displaystyle\text{H}}{\text{C}}=\overset{\displaystyle\text{H}}{\text{C}}\text{H} + \text{HCl} \longrightarrow \text{H}_3\text{C}-\underset{\displaystyle\text{Cl}}{\overset{\displaystyle\text{H}}{\text{C}}}-\text{CH}_3$$

$$\text{H}_3\text{C}-\overset{\displaystyle\text{H}}{\text{C}}=\overset{\displaystyle\text{H}}{\text{C}}-\text{CH}_3 + \text{H}_2 \overset{\text{Pt}}{\longrightarrow} \text{CH}_3\text{CH}_2\text{CH}_2\text{CH}_3$$

$$\text{H}_3\text{C}-\underset{\displaystyle\text{H}}{\overset{\displaystyle\text{CH}_3}{\text{C}}}=\text{CH} + \text{H}_2\text{O} \overset{\text{H}_2\text{SO}_4}{\longrightarrow} \text{H}_3\text{C}-\underset{\displaystyle\text{CH}_3}{\overset{\displaystyle\text{CH}_3}{\text{C}}}-\text{OH}$$

The last reaction above is the reverse of the dehydration reaction outlined in the last section, and the mechanism of the reaction is—in a time sequence—

the reverse of the dehydration mechanism:

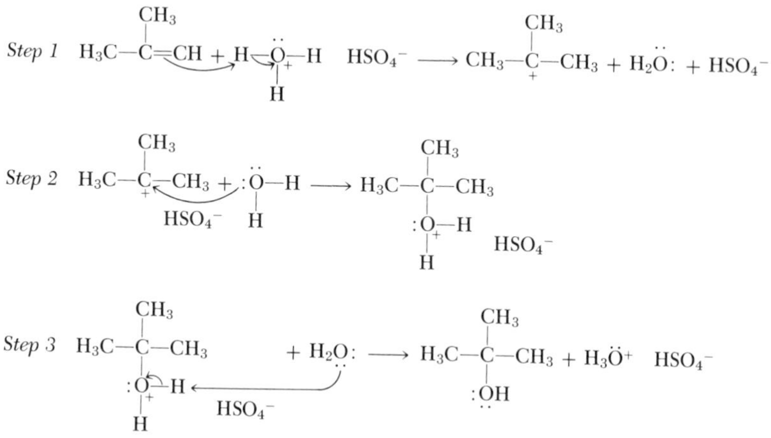

Step 1 $H_3C-\overset{\overset{\displaystyle CH_3}{|}}{C}=CH + H-\overset{\overset{\displaystyle ..}{\underset{\underset{\displaystyle H}{+}}{O}}}{}-H \quad HSO_4^- \longrightarrow CH_3-\overset{\overset{\displaystyle CH_3}{|}}{\underset{+}{C}}-CH_3 + H_2\overset{..}{O}: \; + HSO_4^-$

Step 2 $H_3C-\overset{\overset{\displaystyle CH_3}{|}}{\underset{\underset{\displaystyle HSO_4^-}{+}}{C}}-CH_3 + :\overset{..}{\underset{\underset{\displaystyle H}{}}{O}}-H \longrightarrow H_3C-\overset{\overset{\displaystyle CH_3}{|}}{\underset{\underset{\displaystyle :\overset{..}{\underset{\underset{\displaystyle H}{+}}{O}}-H \quad HSO_4^-}{|}}{C}}-CH_3$

Step 3 $H_3C-\overset{\overset{\displaystyle CH_3}{|}}{\underset{\underset{\displaystyle :\overset{..}{\underset{\underset{\displaystyle H}{+}}{O}}-H \quad HSO_4^-}{|}}{C}}-CH_3 \quad + H_2\overset{..}{O}: \longrightarrow H_3C-\overset{\overset{\displaystyle CH_3}{|}}{\underset{\underset{\displaystyle :\overset{..}{O}H}{|}}{C}}-CH_3 + H_3\overset{..}{O}^+ \quad HSO_4^-$

The hydration-dehydration reactions are reversible:

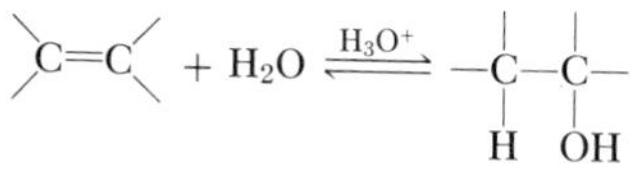

$$\overset{}{\underset{}{C}}=\overset{}{\underset{}{C} + H_2O \underset{}{\overset{H_3O^+}{\rightleftharpoons}} -\overset{|}{\underset{\underset{\displaystyle H}{|}}{C}}-\overset{|}{\underset{\underset{\displaystyle OH}{|}}{C}}-}$$

and strictly speaking, both alkene and hydroxyalkane are present at equilibrium. However, large amounts of water favor the hydration reaction whereas high concentrations of acid (for example, concentrated sulfuric acid) and high temperatures favor the dehydration reaction. Depending on the conditions, it is quite possible to obtain 70- to 100-percent yields of the hydroxyalkanes, or 70- to 100-percent yields of the alkenes at equilibrium.

The reactions of alkenes with the hydrohalic acids proceed by a similar mechanism.

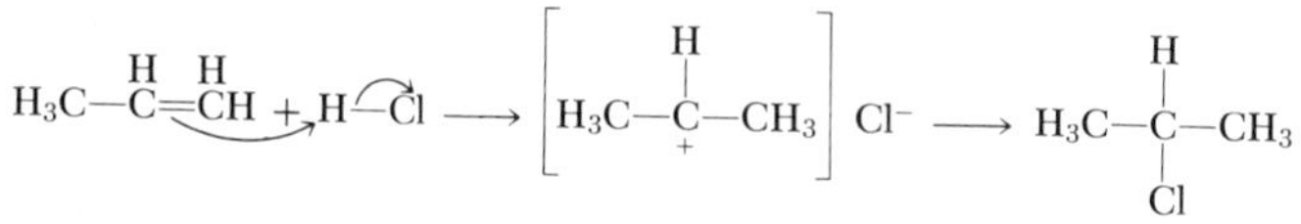

$$H_3C-\overset{\overset{\displaystyle H}{|}}{C}=\overset{\overset{\displaystyle H}{|}}{C}H + H-Cl \longrightarrow \left[H_3C-\overset{\overset{\displaystyle H}{|}}{\underset{+}{C}}-CH_3 \right] Cl^- \longrightarrow H_3C-\overset{\overset{\displaystyle H}{|}}{\underset{\underset{\displaystyle Cl}{|}}{C}}-CH_3$$

The product of this reaction is exclusively 2-chloropropane. The reason why no 1-chloropropane is formed or why no 2-methyl-1-propanol was formed in the previous example will be discussed later in the section on aromatic compounds.

Electrophiles and nucleophiles   The majority of alkene addition reactions are acid-catalyzed, or they involve the addition of acids; in each case, the first step is the attack of a proton or positive ion—the *electrophile*, or electron-seeking reagent—on the π electrons of the double bond; reactions of this type are called *electrophilic* reactions. As we shall see in the next chapter, many reactions involve bases and negative ions—*nucleophiles,* or electron donors—in the initial step; reactions of this type are called *nucleophilic* reactions. Consideration of the nature of such reactions led G. N. Lewis, an American chemist, to propose a definition of acids and bases more general than the Brønsted definitions. According to Lewis, acids are electron acceptors and bases are electron donors; that is, in the following example, the carbonium ion is a *Lewis acid* and ammonia is a *Lewis base*:

$$
\begin{array}{ccc}
\text{H} & \text{CH}_3 & \text{H} \quad \text{CH}_3 \\
| & | & | \quad | \\
\text{H}-\text{N}: + \text{CH}_3-\text{C}^+ & \longrightarrow & \text{H}-\text{N}^{\pm}-\text{C}-\text{CH}_3 \\
| & | & | \quad | \\
\text{H} & \text{CH}_3 & \text{H} \quad \text{CH}_3
\end{array}
$$

The analogy of this reaction to the neutralization of ammonia by hydrochloric acid is obvious:

$$
\begin{array}{cc}
\text{H} & \text{H} \\
| & | \\
\text{H}-\text{N}: + \text{H}_3\text{O}^+\text{Cl}^- \longrightarrow & \text{H}-\text{N}^{\pm}-\text{H} \quad \text{Cl}^- + \text{H}_2\text{O} \\
| & | \\
\text{H} & \text{H}
\end{array}
$$

Polymerization   The formation of giant molecules by the addition of one molecule to another is called *polymerization,* and the product is a *polymer.* We have seen in Chapter 2 that silicon dioxide (quartz or sand) is an example of an inorganic polymer. An organic polymer is readily formed when ethylene is heated with certain compounds (R—R) which dissociate to give free radicals, and polymerization occurs by a free-radical chain reaction similar to that involved in the chlorination of the alkanes.

Initiation $\text{R}-\text{R} \longrightarrow 2\,\text{R}\cdot$

Propagation
$$
\left[
\begin{array}{l}
\text{R}\cdot + \text{CH}_2{=}\text{CH}_2 \longrightarrow \text{R}-\text{CH}_2-\text{CH}_2\cdot \\
\text{R}-\text{CH}_2-\text{CH}_2\cdot + \text{CH}_2{=}\text{CH}_2 \longrightarrow \text{R}-\text{CH}_2-\text{CH}_2-\text{CH}_2-\text{CH}_2\cdot \\
\text{R}-\text{CH}_2-\text{CH}_2-\text{CH}_2-\text{CH}_2\cdot + \text{CH}_2{=}\text{CH}_2 \longrightarrow \text{R}-(\text{CH}_2)_n\cdot
\end{array}
\right.
$$

Termination
$$
\left[
\begin{array}{l}
\text{R}-(\text{CH}_2)_n\cdot + \text{R}\cdot \longrightarrow \text{R}-(\text{CH}_2)_n-\text{R} \\
2\,\text{R}-(\text{CH}_2)_n\cdot \longrightarrow \text{R}-(\text{CH}_2)_{2n}-\text{R}
\end{array}
\right.
$$

The polymer formed in this way, called *polyethylene,* is of considerable commercial importance. Since n is a very large number (1,000 or larger), the polymer molecule is essentially a very long-chain alkane. Natural rubber is a polymer of this type in which the repeat unit (or *monomer*) is 2-methylbutadiene (also called *isoprene*).

Polymers are also formed by the splitting out of water between adjacent monomer units. Nylon and Dacron are commercial polymers made in this way. Many biological materials, such as cellulose, glycogen, proteins, and nucleic acids, are also polymers of essentially this type.

The alkynes (often called *acetylenes*) are hydrocarbons containing one or more triple bonds. Their names are obtained by substitution of the suffix *-yne* for the alkane suffix *-ane;* the simplest member (ethyne) is usually called acetylene, however. Representative members of this class of compounds are given below.

H—C≡C—H   H₃C—C≡C—CH₃

Acetylene   *2-Butyne*

H₃C—C≡C—H   H—C≡C—C≡C—H

Propyne   *Butadiyne*

The structure of the alkynes is derived from still another hybridization of carbon.

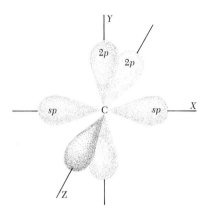

Electron distribution in an excited carbon atom | Two *sp* hybrid orbitals | *sp* hybridized carbon atom | Electron distribution in an *sp* hybridized carbon atom

Figure 3.21   *An sp hybridized carbon atom.*

The bond angle between *sp* orbitals is 180° and the remaining *p* orbitals are found at 90° both to one another <u>and</u> to the *sp* axis (Figure 3.21). In the acelylene molecule, one *sp* orbital is used to form the bond to hydrogen, the other is used to form a σ bond to the second carbon atom, and the *p* orbitals are used to form two π orbitals (Figure 3.22).

Figure 3.22   *The π electron overlap of two sp hybridized carbon atoms.*

There is further overlapping between the π bonds, with the result that the four lobes of the π orbitals form new orbitals that are cylindrical and symmetric about the C—C bond axis (Figure 3.23). The molecular orbital pictures are inconvenient to use, however, so the triple bond is usually indicated by three dashes representing the three electron pairs of the bond. Because of the *sp* hybridization, acetylene is a linear molecule and the four carbon atoms in 2-butyne and butadiyne, for example, fall on a single line, the line of the bond axes.

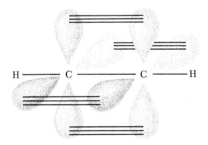

Figure 3.23   *Cylindrical π electron cloud of the acetylene molecule.*

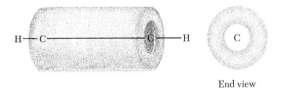

End view

As might be expected from the structures of the double and triple bonds, the chemical reactions and syntheses of the alkynes are quite similar to those of the alkenes.

$$H_3C—C\equiv C—H + 2\ HBr \longrightarrow H_3C—CBr_2—CH_3$$
$$H_3C—C\equiv C—CH_3 + 2\ Br_2 \longrightarrow H_3C—CBr_2—CBr_2—CH_3$$
$$CH_3CBr_2CH_3 + 2\ KOH \longrightarrow H_3C—C\equiv C—H - 2\ K^+Br^- + 2\ H_2O$$

A number of interesting acetylenes have been isolated recently from various wild flowers, principally from the *Compositae*. A surprising example, trideca-1-ene-3,5,7,9,11-pentayne, is given here, with some attempt to indicate the rodlike shape of the molecule:

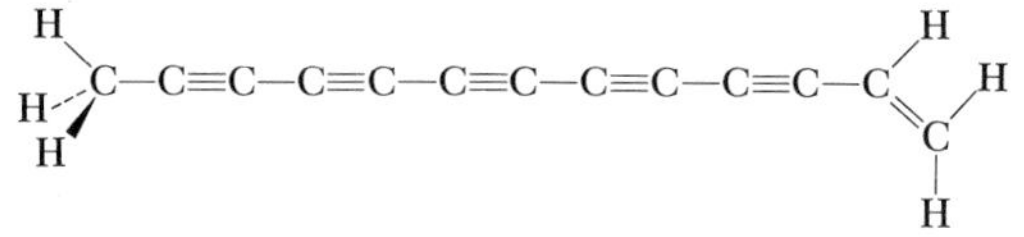

Trideca-1-ene-3,5,7,9,11-pentayne

It is believed that the mechanism for triple-bond production in biological systems is similar to the chemical mechanism outlined in the section on the synthesis of alkenes:

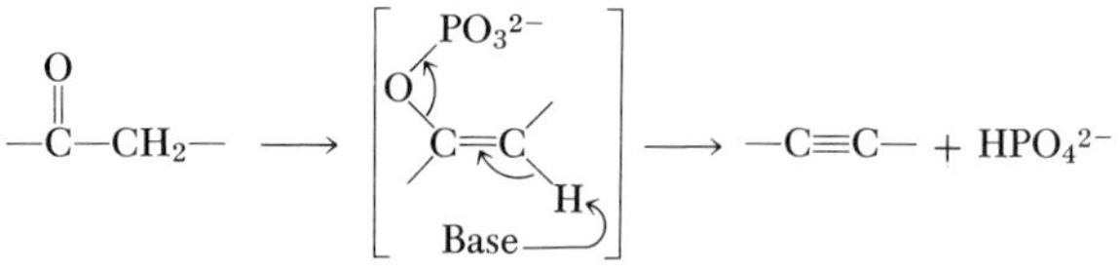

The nature of the phosphate bond will be discussed in Chapter 5.

Aromatic hydrocarbons are unsaturated hydrocarbons that resemble benzene (C_6H_6) in behavior. The name of this class of compounds was assigned in the nineteenth century from the observation that many members had fragrant odors. The simplest member of the series, benzene, is a cyclic molecule containing a six-membered ring and three double bonds.

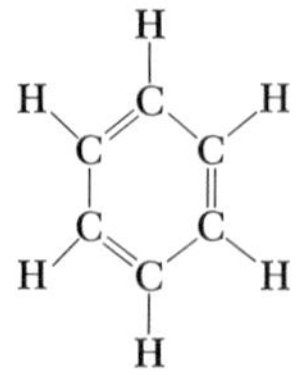

This structure was first proposed by the German chemist August Kekulé in 1865, and it successfully accounted for the chemistry of benzene known at that time. It soon became apparent, however, that the structure was inadequate since it suggested that benzene was simply a cyclic olefin, whereas the chemical reactions of benzene proved to be quite different from those of the olefins. Refinements in the structure of benzene, which successfully account for its properties, have been made in light of modern structural theory, and these will be outlined in a later section; for the time being, we shall continue to use a hexagon with three double bonds as the symbol for benzene. Examples of various aromatic hydrocarbons are given in Figure 3.24, along with their systematic names (the hydrogens attached to the ring are usually omitted from the graphic formulas of aromatic molecules).

The monocyclic aromatic hydrocarbons are usually named as derivatives of benzene whereas the higher members are named as derivatives of a particular ring structure (Figure 3.24b). It is of some interest that 10-methyl-1, 2-benzanthracene (Figure 3.24b) and related compounds are *carcinogenic;* that is, they are able to initiate the formation of cancerous growths in animal tissues.

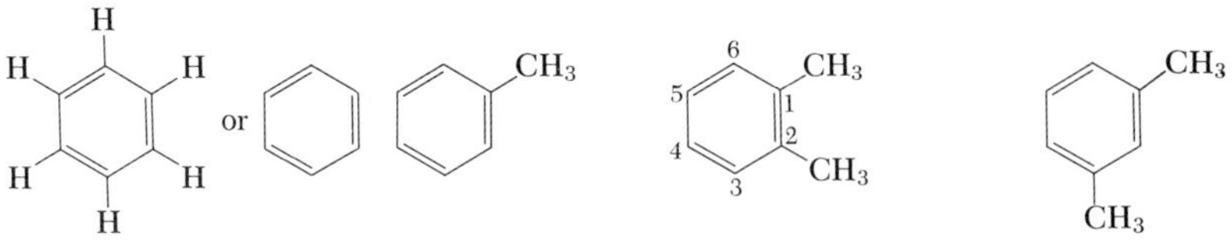

Benzene   Methylbenzene   1,2-Dimethylbenzene   1,3-Dimethylbenzene
           (or toluene)   (or ortho-dimethyl-   (or meta-dimethyl-
                            benzene)   benzene)

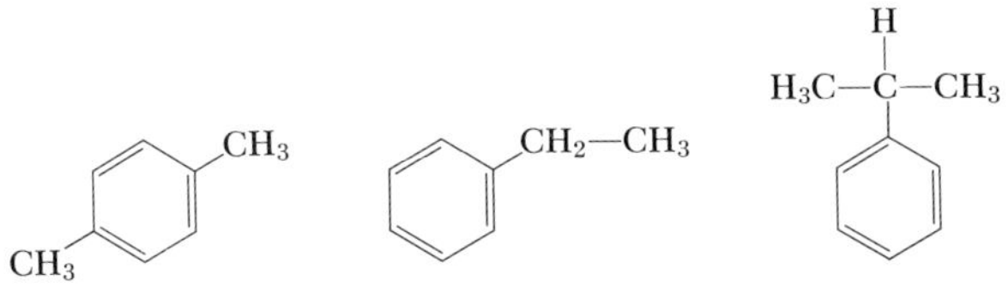

Figure 3.24 (a) Benzene and its alkyl derivatives. (b) Polycyclic aromatic compounds.

1,4-Dimethylbenzene   Ethylbenzene   2-Phenylpropane
(or para-dimethylbenzene)   (or 2-propylbenzene)

a

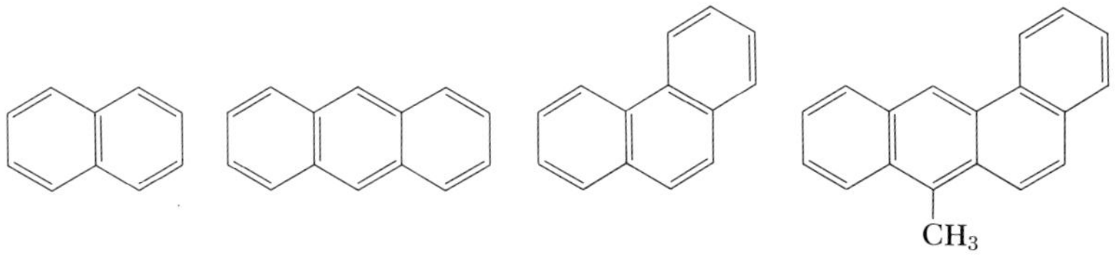

Naphthalene   Anthracene   Phenanthrene   10-Methyl-1,2-
                                             benzanthracene

b

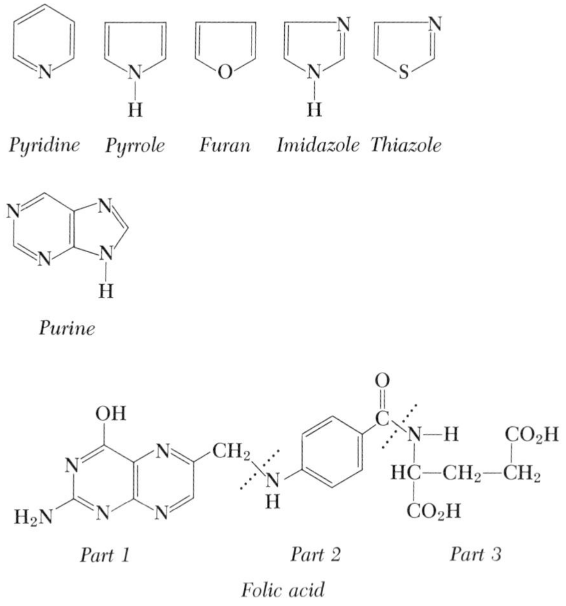

Pyridine   Pyrrole   Furan   Imidazole   Thiazole

Purine

| Part 1 | Part 2 | Part 3 |

Figure 3.25   *Heterocyclic compounds.*   Folic acid

Although they are not hydrocarbons, the heterocyclic compounds should be mentioned at this point since some members of this class are very similar to the aromatic hydrocarbons. The *heterocyclics* are cyclic unsaturated compounds that contain one or more atoms of the elements in the right-hand portion of the periodic table. Examples of heterocyclic compounds are given in Figure 3.25.

Folic acid, one of the B vitamins, is a complex compound made up of the heterocyclic compound 2-amino-4-hydroxy-6-methylpteridine (part 1), the aromatic acid *para*-aminobenzoic acid (part 2), and the amino acid glutamic acid (part 3). Its general role in the body and its relation to the mode of action of the antibiotic sulfanilamide will be outlined in Chapter 6. Still other heterocyclic compounds, in particular cytosine (p. 141), thymine, uracil, guanine, and adenine—components of the nucleic acids (DNA and RNA)—are discussed in a different volume in this series.[*]

The chemistry of the heterocyclic compounds cannot be outlined in this brief volume. However, the characteristic reactions are those of the aromatic ring (see below, Examples *A* and *C*), and those of the oxygen, nitrogen, or sulfur atoms that are present in the heterocyclic compounds (Example *B*).

[*] W. D. McElroy, *Cell Physiology and Biochemistry*, 3rd ed. (Englewood Cliffs, N. J.: Prentice-Hall, Inc., 1969).

Example A:

Thiophene + Br$_2$ $\longrightarrow$ (2-bromothiophene) + HBr

Example B:

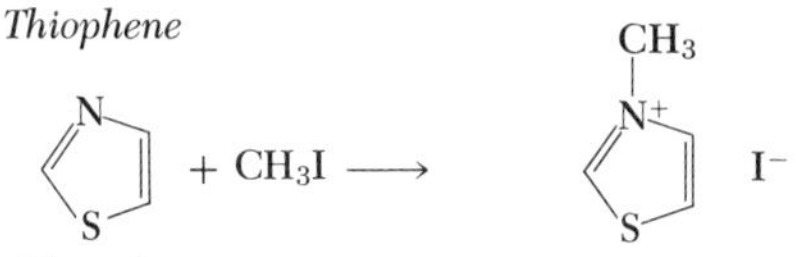

Thiazole + CH$_3$I $\longrightarrow$ (N-methylthiazolium) I$^-$

Example C:

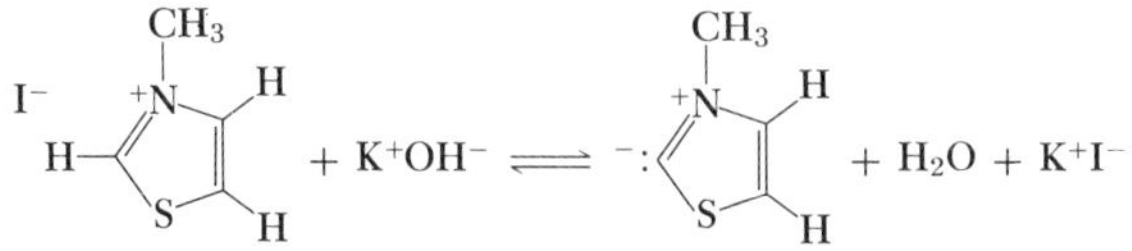

N-Methylthiazolium iodide + K$^+$OH$^-$ $\rightleftharpoons$ (ylide) + H$_2$O + K$^+$I$^-$

General reactions of these types will be covered in Chapters 3 to 6.

REACTIONS OF AROMATIC COMPOUNDS   The characteristic reactions of aromatic compounds are substitutions; that is, reactions in which some group is substituted for a hydrogen atom on the ring. Examples are given below.

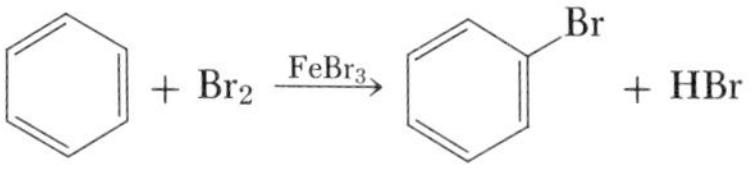

+ Br$_2$ $\xrightarrow{\text{FeBr}_3}$ (Br substituted benzene) + HBr

Bromobenzene

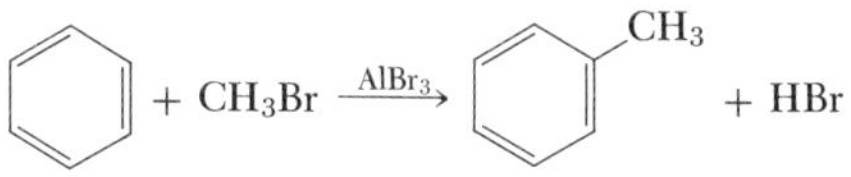

+ CH$_3$Br $\xrightarrow{\text{AlBr}_3}$ (CH$_3$ substituted benzene) + HBr

Toluene

+ 3 HONO$_2$ $\xrightarrow{\text{H}_2\text{SO}_4}$ (trinitrotoluene) + 3 H$_2$O

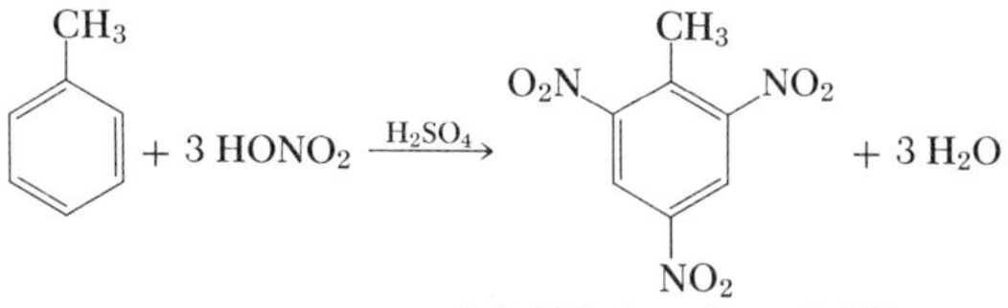

2,4,6-Trinitrotoluene (TNT)

These substitution reactions of benzene are in marked contrast to the addition reactions of the alkenes and alkynes. A noncyclic isomer of benzene, 1,3-hexadiene-5-yne (H—C≡C—CH=CH—CH=CH$_2$), for example, reacts typically by addition; furthermore, linear compounds of this type are far less stable than benzene. These differences in properties are satisfactorily accounted for by a theory developed during the early part of this century, the theory of resonance, which has proved to be of great value in interpreting the stability of compounds, the nature of the chemical reactions undergone, and the orientation, or direction of chemical attack; we shall now examine this theory.

RESONANCE   Molecules containing only single bonds are satisfactorily represented by formulas in which a dash is used to symbolize an electron-pair bond, as in methane:

$$H:\overset{\overset{\displaystyle H}{\cdot\cdot}}{\underset{\displaystyle H}{C}}:H \;=\; H-\overset{\displaystyle H}{\underset{\displaystyle H}{C}}-H$$

The electrons in bonds of this type are largely localized between the atoms forming the bond. This is essentially true also for molecules containing one double or triple bond, and also for those containing several such bonds if the multiple bonds are separated by a saturated carbon atom, as in 1,4-pentadiene:

$$\underset{\displaystyle H}{\overset{\displaystyle H}{{}}}C=\overset{\displaystyle H}{C}-\overset{\displaystyle H}{\underset{\displaystyle H}{C}}-\overset{\displaystyle H}{C}=C\overset{\displaystyle H}{\underset{\displaystyle H}{{}}}$$

The symbolism breaks down, however, for molecules that contain two or more multiple bonds attached directly to one another as in butadiene and benzene, and for molecules or ions that contain atoms with unshared electrons attached directly to a double bond. The π electrons in molecules or ions of this type are not localized; they are spread out over all the atoms involved in the multiple bonding and the atoms bearing unshared electrons. For example, we can readily write a structure for the carbonate ion $(CO_3{}^{2-})$ in which each atom has an octet of electrons:

$$\underset{}{\overset{:\ddot{O}:}{\underset{{}^{-}:\ddot{O}\quad\ddot{O}:^{-}}{\overset{\|}{C}}}}$$

Two oxygen atoms are connected to carbon by σ bonds, and the third is connected by both σ and π bonds. There is no reason why one oxygen atom should differ from the other two, however, so we must consider two more structures:

$$\underset{{}^{-}:\ddot{O}\quad\ddot{O}}{\overset{:\ddot{O}:^{-}}{C}} \qquad and \qquad \underset{:\ddot{O}\quad\ddot{O}:^{-}}{\overset{:\ddot{O}:^{-}}{C}}$$

Is it possible that all the oxygen atoms are bonded to carbon in the same way? X-ray diffraction studies show that this is in fact the case; the three C—O bond lengths in the carbonate ion are identical.

Compounds or ions of this type, then, cannot be represented satisfactorily by a single valence-bond structure; they are represented instead by a series of structures connected by double-headed arrows.

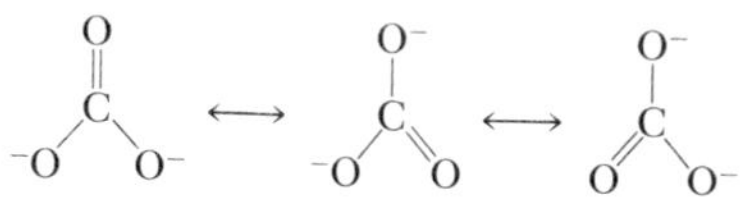

The "true structure" of the compound is an average of the structures drawn. Molecules or ions of this type are called *resonance hybrids*, and the individual structures that can be drawn are called *contributing* structures.

The representation of resonance hybrids in this way is an attempt to show that the unshared electrons in the CO_3^{2-}, for example, are not localized on two of the oxygen atoms and that the π electrons are not localized between the carbon atom and one oxygen atom (as shown in any contributing structure), but that these electrons are delocalized, or spread out, over all the atoms with available p orbitals. A second way to represent the resonance hybrids involves the use of dotted lines to indicate the π-p system over which the electrons are delocalized:

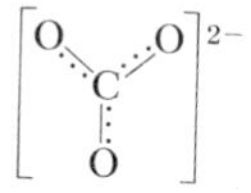

This type of formula more clearly indicates the symmetry of the system.

Similarly, if a proton is removed from a propene molecule

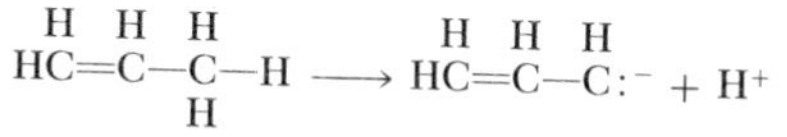

the electron pair is not localized on one carbon atom, but is delocalized over the π network:

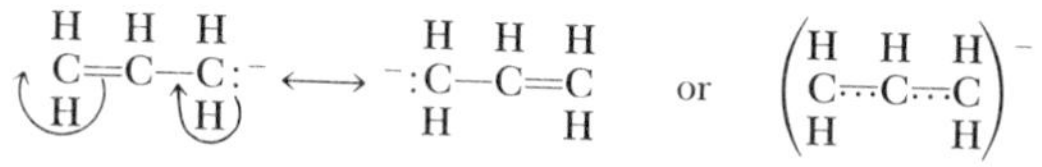

This delocalization can be represented by the interaction of the π and p orbitals involved, as in Figure 3.26.

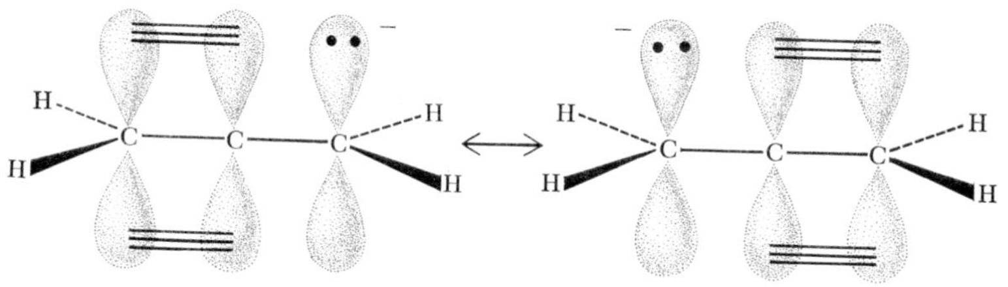

Figure 3.26 ***Delocalization of electrons in the negative ion of propene.***

The resonance of benzene　　The resonance concept is required for an understanding of the structures of aromatic compounds. Benzene, for example, is not represented well by a single formula:

These formulas stand for "cyclohexatriene," a molecule that should show the reactions of an alkene. Neither formula represents benzene, then, which as we have seen from its chemical reactions is not an alkene. Instead, we represent benzene by a pair of resonance contributors, or by a single structure with dotted lines, which is supposed to be an average of the two extreme forms:

In this way we imply delocalization of the π electrons.

　　The delocalization of electrons in benzene can be illustrated more clearly through the use of molecular orbitals. The six carbon atoms of benzene are sp^2 hybrids, and a six-membered ring may be formed by using two sp^2 orbitals from each atom to form bonds to two neighboring carbon atoms; the remaining sp^2 orbital forms a bond to hydrogen, as at the bottom of the page.

　　If adjacent p orbitals are now allowed to form ordinary two-atom π orbitals, the resulting two equivalent structures (Figures 3.27a and 3.27b) would correspond to our individual resonance contributors. However, a molecular orbital

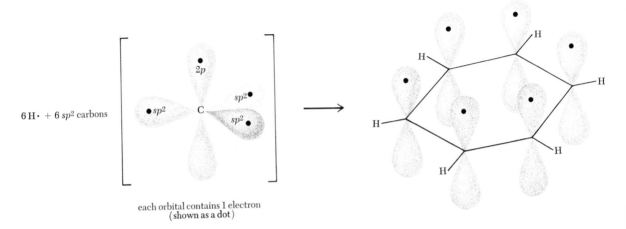

$6\,H\cdot\ +\ 6\ sp^2$ carbons

each orbital contains 1 electron
(shown as a dot)

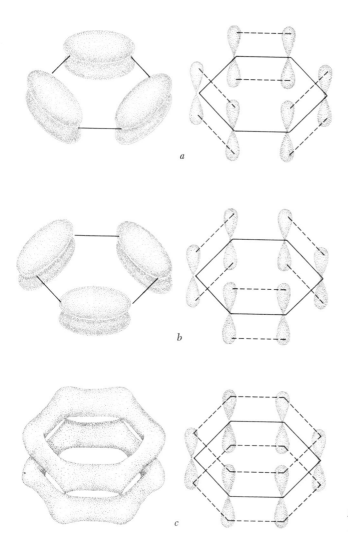

Figure 3.27 **Resonance** *in benzene.* [From C. R. Noller, *J. Chem. Ed.*, 27 (1950), 505.]

can be constructed in which overlap occurs between all of the p orbitals, allowing complete delocalization of the π electrons (Figure 3.27c). This orbital contains two of the π electrons of benzene; the other four are placed in two other similar orbitals. This physical picture of the molecule satisfactorily accounts for the facts that are known about the structure of benzene. The benzene molecule is planar and highly symmetrical; the six C—H bonds are identical in length, the six carbon–carbon bonds are identical in length, and all the bond angles are the same (120°). It is the closed ring of π electrons that accounts for the special properties of benzene.

ORGANIC CHEMISTRY: THE HYDROCARBONS

The benzene molecule contains a closed ring of six π electrons. Organic chemists have studied many other monocyclic structures with π electrons, and it is interesting to note that only those compounds with a closed ring of $4n + 2\pi$ electrons have aromatic properties (the n in "$4n + 2$" can be any integer). This rule predicts that aromatic compounds can be prepared with what might be called "magic" numbers of 2, 6, 10, 14, 18, . . . π electrons, corresponding to values of n of 0, 1, 2, 3, 4, ... Benzene corresponds to a magic number of 6, and recently new compounds have been synthesized corresponding to magic numbers of 2, 6, and 18.

The number of isomers of the substituted benzenes is also in accord with the resonance theory of benzene. If the electrons were not delocalized in the aromatic ring, we would expect two isomers of 1,2-dimethylbenzene:

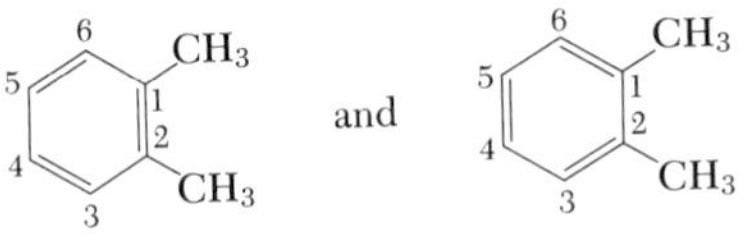

one with a double bond connecting carbon atoms 1 and 2, and the other with a single bond connecting those atoms. But the resonance picture suggests that 1,2-dimethylbenzene should exist in a single form.

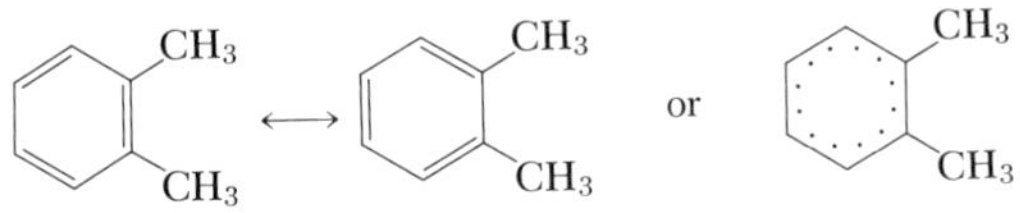

Many careful attempts have been made to determine the homogeneity of this compound and related 1,2-disubstituted benzenes, and in every case, the resonance theory has been upheld; that is, only one form of these compounds has been found.

Resonance energy   A resonance hybrid is more stable (that is, it contains less energy) than would be expected from the contributing structures. This difference is the *resonance energy* of the compound. The resonance energy may be approximated by measuring the energy released during a chemical reaction. For example, the reaction of cyclohexene with hydrogen over a nickel catalyst yields cyclohexane, and 28,000 cal of energy are released in the form of heat (ΔH).

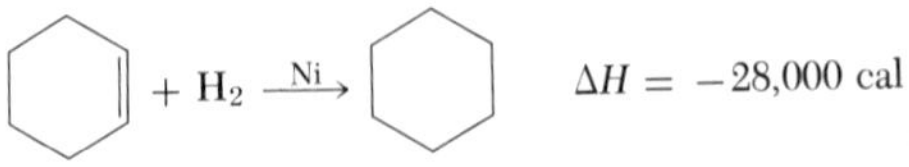

This value of ΔH is also found for the hydrogenation of ethylene, propene, and other alkenes with only one double bond. Furthermore, the hydrogenation of 1,4-cyclohexadiene liberates 56,000 cal:

$$\text{[structure]} + 2\ H_2 \xrightarrow{\ Ni\ } \text{[structure]} \qquad \Delta H = -56,000\ \text{cal}$$

That is, the same amount of energy is released during the hydrogenation of each double bond. We now assume that our hypothetical resonance contributor to benzene (which we will call "cyclohexatriene," no delocalization of π electrons being permitted) would yield $3(28,000) = 84,000$ cal on hydrogenation.

Hypothetical case $\qquad \text{[structure with H labels]} + 3\ H_2 \xrightarrow{\ Ni\ } \text{[structure]} \qquad \Delta H = -84,000\ \text{cal}$

The hydrogenation of benzene, however, actually yields 48,000 cal or 36,000 cal less than the cyclohexatriene value. This means that benzene contains 36,000 cal less energy than does "cyclohexatriene," or, in effect, that it is more stable by 36,000 cal. This difference in energy (36,000 cal) is the resonance energy

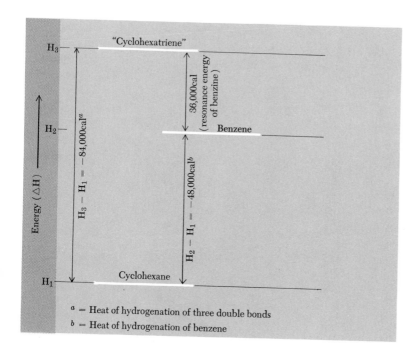

Figure 3.28 *The resonance energy of benzene.*

a = Heat of hydrogenation of three double bonds
b = Heat of hydrogenation of benzene

ORGANIC CHEMISTRY: THE HYDROCARBONS

of benzene (also called the delocalization energy). The relationships just outlined are illustrated in Figure 3.28.

Several analogs of benzene and a few other resonance-stabilized molecules are given in Figure 3.29, along with the principal resonance contributors, and values of the resonance energy. (The curved arrows signify the movement of electron pairs and they lead to the formulas nearest them on the right.) Very often only one resonance contributor is used to represent an aromatic compound, as in Figure 3.24; this is done purely for convenience, and in such cases the structure stands for the resonance-stabilized molecule.

The resonance contributors shown for phenol in Figure 3.29 illustrate that unshared p electrons on atoms attached directly to a double bond or aromatic system are also delocalized over the π electron network. The charges indicated on these resonance contributors are called *formal charges;* they represent the difference in the number of electrons "possessed" by the bonded atom and the number in the valence shell of the free atom. In nitromethane,

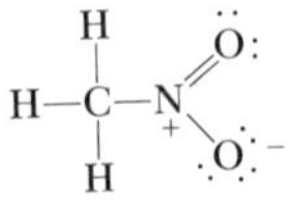

Figure 3.29   Resonance energy and principal resonance contributors for selected compounds.

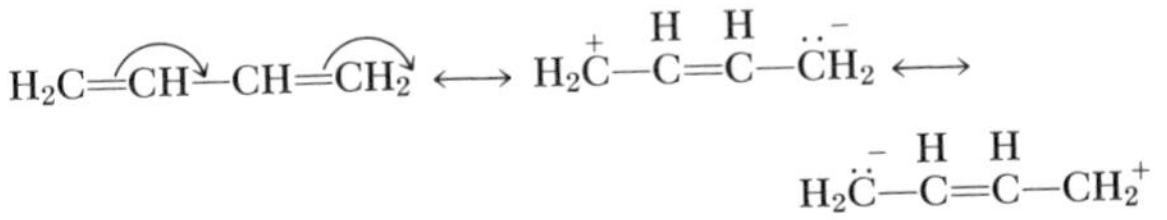

Butadiene *(3,500 cal)*

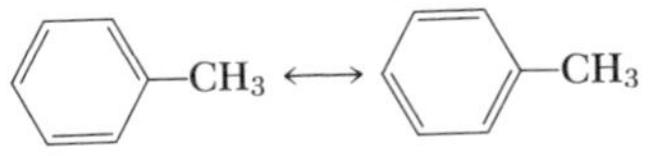

Toluene *(36,000 cal)*

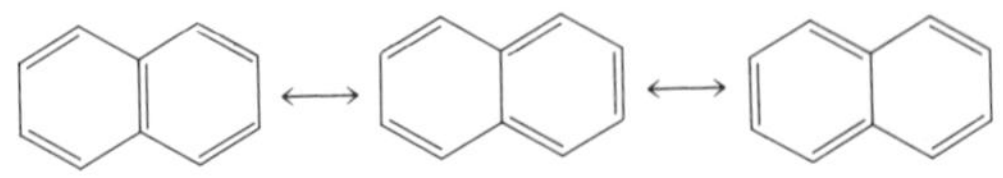

Naphthalene *(61,000 cal)*

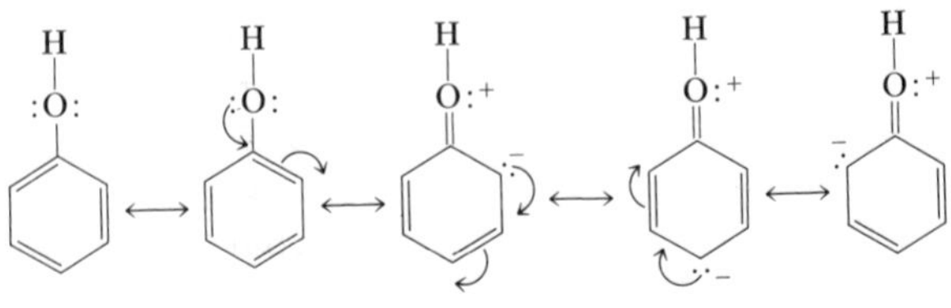

Hydroxybenzene *(phenol) (36,000 cal)*

the singly bonded oxygen atom contains three pairs of unshared electrons (6 e) and it shares one electron pair with nitrogen ($\frac{1}{2}$ of 2 = 1 e). This bonded oxygen atom, therefore, possesses seven electrons, whereas the valence shell of an isolated oxygen atom contains six electrons ($:\ddot{O}:$). The formal charge indicated on the oxygen atom in nitromethane (-1) gives this difference in the number of electrons ($7 - 6 = 1$ e). The formal charge on the nitrogen atom is arrived at in a similar fashion.

Resonance energy and the electrophilic substitution of aromatic compounds The concept of resonance accounts for the fact that substitution rather than addition is the characteristic reaction of aromatic hydrocarbons. Most of the reactions of aromatic compounds are electrophilic substitution reactions; that is, they are reactions in which electrophiles (for example, positively charged ions) attack the ring. Thus, the mechanism for bromination of benzene catalyzed by ferric bromide is

$$FeBr_3 + Br_2 \xrightarrow[1]{step} FeBr_4^- + Br^+$$

The ferric bromide in step 1 is a transition metal compound, and as such it is a Lewis acid (an electron acceptor). Thus, it pulls a negative bromide ion from bromine to give the complex ion $FeBr_4^-$, leaving the species Br^+. The Br^+ ion attacks the π electron system in step 2 to give an adduct with three principal resonance contributors; these contributors show that the positive charge is not localized on the second carbon atom, but that it resides in addition on carbon atoms 4 and 6. Note that the resonance is of the propenyl type (p. 81) and that the positive charge can be placed only on alternate carbon atoms; inspection of the figure will show that the positive charge cannot be placed on carbon atoms 3 and 5. The loss of a proton by the carbonium ion in step 3 gives the fully aromatic, resonance-stabilized bromobenzene as the product. If instead of the elimination of a proton a bromide ion were added to the carbonium ion, the product would be a simple cyclohexadiene with very little resonance energy. It is the regaining of a resonance-stabilized system by the loss of a proton that leads to substitution in the aromatic series. In summary, then, resonance-stabilized alkenes or aromatic compounds react by electrophilic substitution reactions, whereas alkenes with little or no resonance energy react by electrophilic addition reactions. Other electrophiles that are capable of substitution reactions with aromatic compounds are NO_2^+ (from nitric acid: $HONO_2 + H^+ \longrightarrow H_2O + NO_2^+$), carbonium ions (from $RCl + AlCl_3 \longrightarrow R^+ + AlCl_4^-$, or $ROH + H_2SO_4$, or alkenes + acids), and $R{-}C^+{=}O$ (from carboxylic acids + H^+); see page 79 for examples of such reactions.

ORIENTATION IN AROMATIC SUBSTITUTION The nitration of toluene under mild conditions yields a mixture of *ortho-* and *para*-nitrotoluenes:

Toluene *ortho-Nitrotoluene* *para-Nitrotoluene*

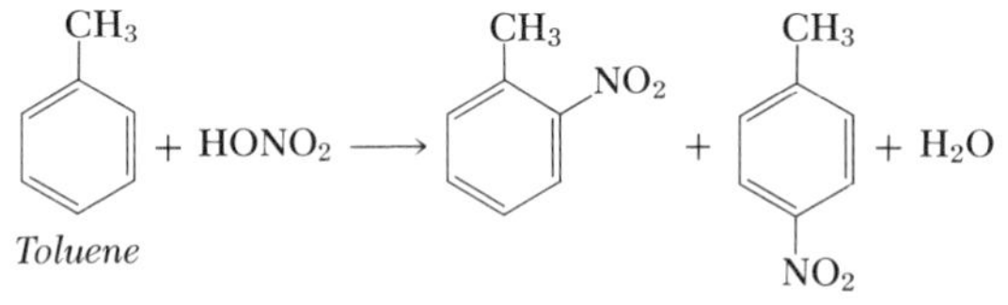

whereas the nitration of nitrobenzene yields principally *meta*-dinitrobenzene;

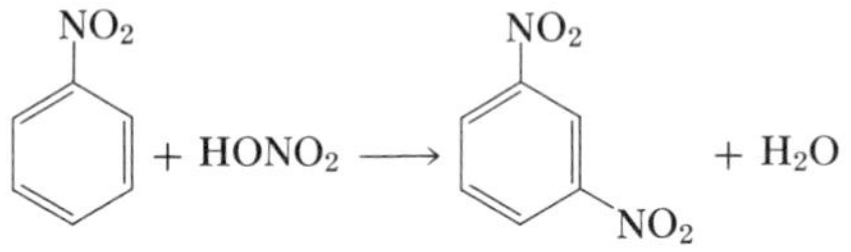

This difference in orientation is determined by the nature of the substituent (or group) on the benzene ring. How the substituent influences the orientation in reactions of this type is related to the explanation of why the addition of hydrogen chloride to propene yields 2-chloropropane and not the 1-chloropropane isomer (see p. 73).

$$H_3C{-}\overset{H}{C}{=}\overset{H}{C}H + HCl \begin{cases} \not\longrightarrow CH_3CH_2CH_2Cl \\ \longrightarrow CH_3CHClCH_3 \end{cases}$$

We shall discuss the question of orientation in aliphatic (nonaromatic) systems first.

Orientation in carbonium ion reactions The carbon atom to which the chlorine becomes bonded is, in reality, determined by the type of carbonium ion formed in the first stage of the reaction, the first stage being the attack of a proton on the π electrons.

$$CH_3{-}\overset{H}{C}{=}\overset{H}{C}H + H^+ \begin{cases} \not\longrightarrow CH_3CH_2CH_2^+ \\ \longrightarrow CH_3\underset{+}{C}HCH_3 \xrightarrow{Cl^-} CH_3CHClCH_3 \end{cases}$$

The carbonium ions formed in reactions of this type are, invariably, the most highly substituted ones. We distinguish three types of carbonium ions by the number of alkyl groups attached to the electron-deficient carbon atom:

$$
\text{Primary carbonium ions:}\quad
\underset{H}{\overset{H}{HC^+}},\quad
\underset{H\ H}{\overset{H\ H}{HC-C^+}},\quad
\text{(phenyl)}-\underset{H}{\overset{H}{C^+}},\ \text{etc.}
$$

$$
\text{Secondary carbonium ions:}\quad
\underset{H\ +\ H}{\overset{H\ H\ H}{HC-C-CH}},\quad
\text{(phenyl)}-\underset{+\ H}{\overset{H\ H}{C-CH}},\ \text{etc.}
$$

$$
\text{Tertiary carbonium ions:}\quad
\underset{HCH}{\overset{HCH}{\underset{H}{\overset{H}{HC-C^+}}}},\quad
\text{(phenyl)}-\underset{HCH}{\overset{HCH}{C^+}},\ \text{etc.}
$$

Alkyl groups are electron-releasing groups (relative to hydrogen) and therefore alkyl groups stabilize (or lower the energy of) carbonium ions. The stability order of carbonium ions is: tertiary > secondary > primary; the ease of forming these carbonium ions also decreases in this order. To return to the addition of hydrogen chloride to propene, the secondary propyl carbonium ion is more stable than the primary ion, and therefore the secondary ion is formed from the addition of the proton to propene. The addition of chloride ion, leading to 2-chloropropane, then terminates the reaction. For similar reasons, the reaction of benzene with propene (catalyzed by sulfuric acid) leads to 2-phenylpropane and not 1-phenylpropane.

$$
H_3C-CH{=}CH_2 \xrightarrow{H^+} H_3C-\underset{+}{\overset{H}{C}}-CH_3 + \text{(benzene)} \xrightarrow{-H^+} \text{(phenyl)}-\underset{H}{\overset{CH_3}{C}}-CH_3
$$

Electron-releasing groups and the ortho-para substitution of aromatic compounds   When attached to carbonium ions or to double bonds, most atoms and groups containing only single bonds are electron-releasing groups. This is illustrated with the aid of the resonance contributors for three derivatives of benzene in Figure 3.30. Notice that the electron-releasing groups place a partial negative charge on the *ortho* and *para* positions of the ring, but *not* on the *meta* positions. This delocalization of electrons is especially important for groups bearing unshared electrons. The attack of an electrophilic species on a ring containing an electron-releasing substituent occurs preferentially in the *ortho* and *para* positions, because the positive charge on the ring in these cases can be partially neutralized by the electron release of the substituent; in other words, the charge is delocalized over the ring and also over the substituent (Figure 3.31a).

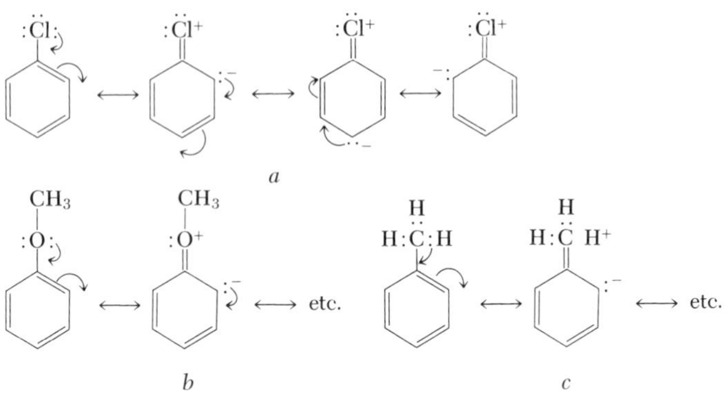

Figure 3.30 *Resonance interaction of electron-releasing groups.*

A similar set of four resonance contributors can be written for substitution in the *para* position. Substitution in the *meta* position, however, leads to an intermediate in which the positive charge on the ring cannot be delocalized over the substituent (Figure 3.31b). Consequently, very few molecules react by *meta* nitration because of the high activation energy of this path. The result of this type of resonance interaction is *ortho-para* substitution in nitration and in other electrophilic substitution reactions of benzene rings containing electron-releasing substituents.

Figure 3.31 **Electron delocalization** (a) during the electrophilic substitution of an aromatic ring ortho to an electron-releasing substituent; (b) during the electrophilic substitution of an aromatic ring in the meta position.

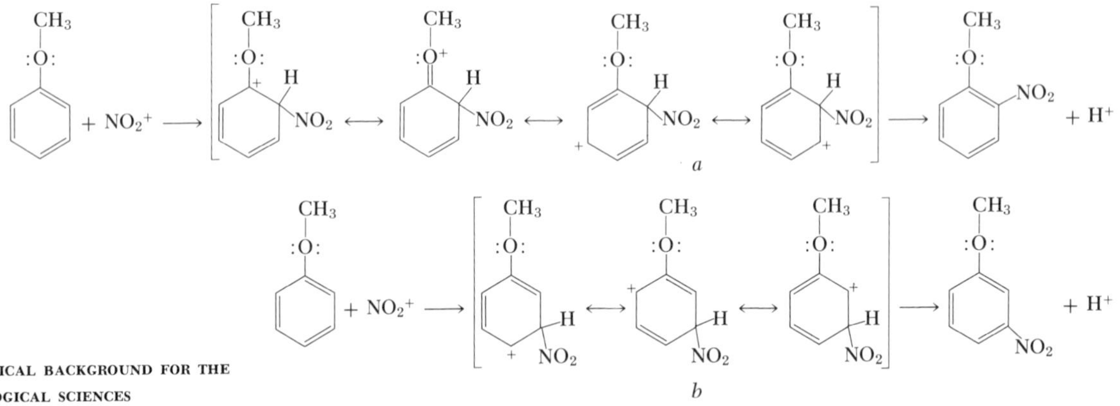

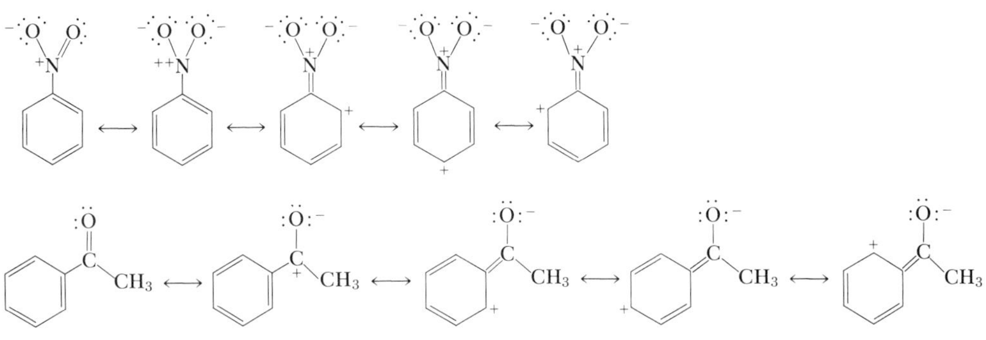

Figure 3.32 **Resonance contributors** *of benzene rings bearing electron-attracting substituents.*

Electron-attracting groups and the meta substitution of aromatic compounds The second major type of substituent is the electron-attracting variety. In general, substituents of this type contain multiple bonds to oxygen or nitrogen, elements that are more electronegative than carbon is. Examples illustrating the resonance interaction of substituents of this type are given in Figure 3.32. Notice that these groups place partial positive charges on the *ortho* and *para* positions. This results in a lower electron density on the benzene ring, and also in electrophilic substitution at a slower rate than in rings containing electron-supplying groups; furthermore, since the electron density is greater in the *meta* position, these groups lead principally to *meta* substitution. Specifically, electrophilic attack at the *ortho* and *para* positions is a relatively high-energy process since resonance involving the positive charge generated by the attack of the electrophile (NO_2^+, for example) places positive charges on adjacent atoms in one of the resonance contributors (Figure 3.33*a*). The repulsion of like charges makes this resonance contributor an improbable one. This means that delocalization effectively occurs over only two positions of the ring (Figure 3.33*a*). In contrast, attack in the *meta* position leads to delocalization over three positions of the ring (Figure 3.33*b*), and in none of the resonance contributors are like charges placed on adjacent atoms.

The classification of sustituents into electron-releasing and electron-attracting groups is fundamental to any study of the reactions a compound will undergo, the rates of such reactions, and the reaction mechanisms involved. We shall apply the principles outlined in this chapter to some concrete examples in the remainder of this volume.

The importance of resonance can be shown in biological reactions as well. For example, benzene, chlorobenzene, and related compounds are toxic to

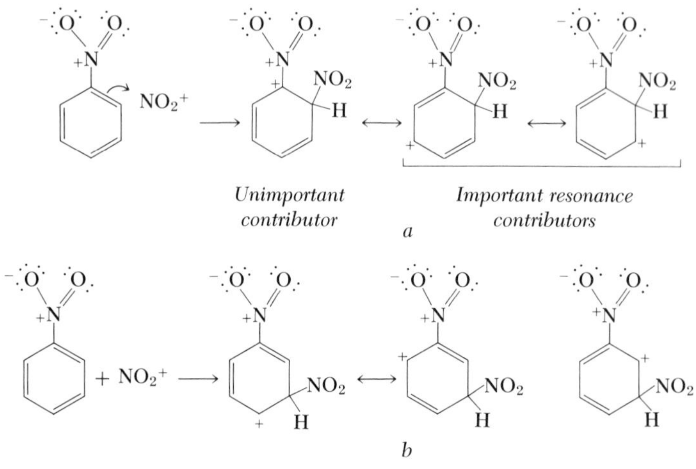

Unimportant
contributor

Important resonance
contributors

a

b

Figure 3.33   (a) **Ortho attack** on a benzene ring
containing an electron-attracting substituent. (b)
Meta attack.

higher organisms, but they are detoxified by the body and are excreted as
mercapturic acids. In the case of chlorobenzene, the *para,* not the *meta,* derivative is formed (as would also be the case in a simple chemical reaction).

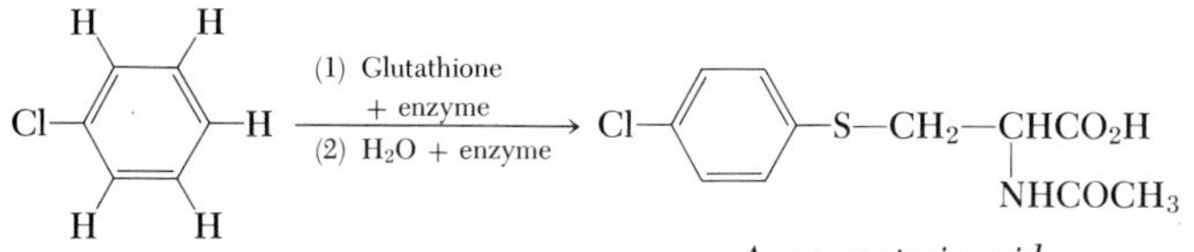

A mercapturic acid

In addition, hydroxylations (as in the conversion of the amino acid phenylalanine into tyrosine) and oxidative coupling of phenols occur in the *para*
position, suggesting that the electron supply is a contributing factor in the
orientation.

4 ORGANIC COMPOUNDS CONTAINING OXYGEN

COMPOUNDS CONTAINING ONLY CARBON, HYDROGEN, AND OXYGEN ARE EXtremely important in organic chemistry. In a structural sense, many of the simple oxygen-containing compounds may be considered as derivatives of the hydrocarbons in which some functional group containing oxygen is attached to a chain or a ring of the hydrocarbon. In this chapter, we shall cover the principal groups of oxygen derivatives in the order of the oxidation level of their functional groups.

ALCOHOLS

The alcohols (R—O—H) are alkyl derivatives of water, or in view of the definition above, are hydroxy derivatives of the hydrocarbons. The systematic names of the alcohols are derived by the substitution of the suffix *-ol* for the suffix *-ane* of the corresponding hydrocarbons; the hydroxy derivatives of benzene, however, are usually referred to as *phenols*. Examples of alcohols and phenols, along with their systematic names and in certain cases their common names are given in Figure 4.1.

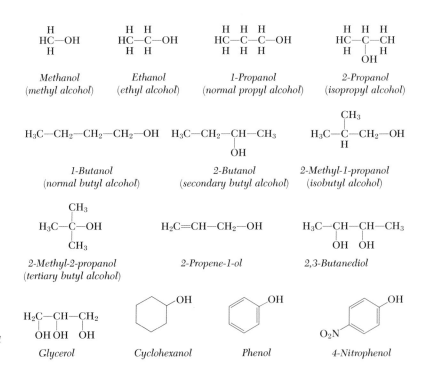

Figure 4.1 Common alcohols and phenols.

The alcohols are widely distributed in nature, although they are very often found chemically bound to other molecules (in the form of esters, acetals, or other derivatives). Ethyl alcohol usually occurs in the free state, however, and as such it is the active principle in beer, wine, and other "alcoholic" beverages; it is commonly prepared by the fermentation of the sugars present in grains, fruits, vegetables, and so on. Many of the naturally occurring alcohols have complex structures, as is exemplified by vitamin A, which is essential for the growth of mammals. Note the relationship of the structure of vitamin A to that of β-carotene (Figure 3.13):

Vitamin A

PROPERTIES OF THE ALCOHOLS   Methanol, ethanol, and the propanols are soluble in water in all proportions, and their physical and chemical properties resemble those of water. Thus, the hydroxyl group dominates both the physical and the chemical properties of the lower alcohols. The alcohols of

Table 4.1 *Boiling points of pairs of compounds with the same or
with approximately the same molecular weight*

COMPOUND	MOLECULAR WEIGHT	BOILING POINT, $^\circ$C
CH_4	16	-161
H_2O	18	100
CH_3CH_3	30	-89
CH_3OH	32	65
H_3C—O—CH_3	46	-24
H_3C—CH_2—OH	46	78

higher molecular weight are either only partly soluble or insoluble in water
because of the long hydrocarbon chains (hydrocarbons are insoluble in water).
In physical properties, therefore, the higher alcohols resemble the alkanes; the
OH groups of these alcohols still dominate the *chemical* reactivity, however,
as we shall see.

The boiling points of water and the alcohols are abnormally high relative
to the boiling points of the hydrocarbons, ethers, and related compounds of
similar molecular weight (Table 4.1). Studies of the alcohols by means of
their absorption of infrared light have provided an explanation for this fact.
The —OH bond is highly polar because of the great electronegativity of
oxygen:

$$R—\overset{..}{\underset{..}{O}}—H \longleftrightarrow R—\overset{..}{\underset{..}{O}}:^-H^+ \text{ (or } R—\overset{\delta^-}{O}—\overset{\delta^+}{H})$$

Thus, hydrogen bonds can be formed between adjacent alcohol molecules
in the same way that hydrogen bonds are formed between water molecules
(p. 24). As a result of this type of weak interaction, alcohol molecules tend
to form aggregates of various sizes in solution:

$$\overset{\delta^-}{\underset{R}{:\overset{..}{O}}}—\overset{\delta^+}{H}---\overset{\delta^-}{\underset{R}{:\overset{..}{O}}}—\overset{\delta^+}{H} \quad \text{and} \quad \underset{R}{O}—H-\left(\underset{R}{O}—H\right)_n---\underset{R}{O}—H$$

The high boiling points result because extra energy is required to break
up these aggregates to allow individual molecules to enter the vapor phase.
Hydrogen bonds are formed between compounds that are weak acids and
weak bases; this limits the phenomenon essentially to compounds containing
NH, OH, and SH bonds. If one of the components were a strong acid or a
strong base, complete proton transfer would occur, as for example in the
reaction of water with strong acids to form the hydronium ion (H_3O^+), or in the
reaction of acids with ammonia to give the ammonium ion (NH_4^+).

Alcohols are prepared by *hydrolysis* (decomposition with water) of the halogen derivatives of the alkanes:

$$RX + H_2O \longrightarrow ROH + HX$$

A similar reaction occurs with hydroxide ion:

$$RX + OH^- \longrightarrow ROH + X^-$$

These substitution reactions proceed by one of two mechanisms, depending on the stability of the carbonium ion derived from the halide, RX. If the carbonium ion is relatively stable, a two-step reaction occurs in hydrolysis:

Step 1 (slow)

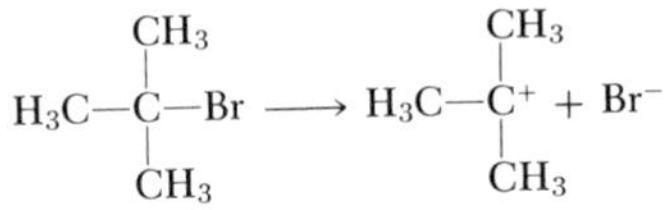

$$H_3C-\underset{\underset{CH_3}{|}}{\overset{\overset{CH_3}{|}}{C}}-Br \longrightarrow H_3C-\underset{\underset{CH_3}{|}}{\overset{\overset{CH_3}{|}}{C}}^+ + Br^-$$

Step 2 (fast)

$$H_3C-\underset{\underset{CH_3}{|}}{\overset{\overset{CH_3}{|}}{C}}^+ + 2\,H_2O \longrightarrow H_3C-\underset{\underset{CH_3}{|}}{\overset{\overset{CH_3}{|}}{C}}-O-H + H_3O^+$$

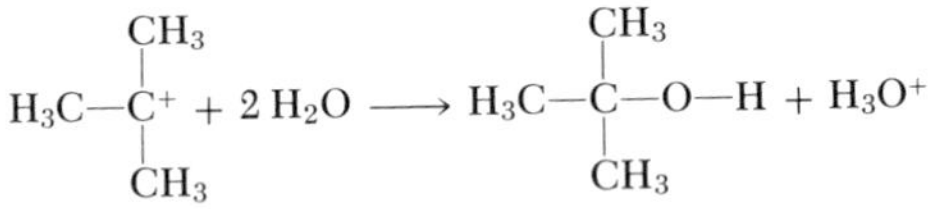

The first step is a slow ionization, giving the tertiary butyl carbonium ion and a bromide ion, and the second step is a fast reaction of the carbonium ion with the nucleophile water, yielding the alcohol. Similarly:

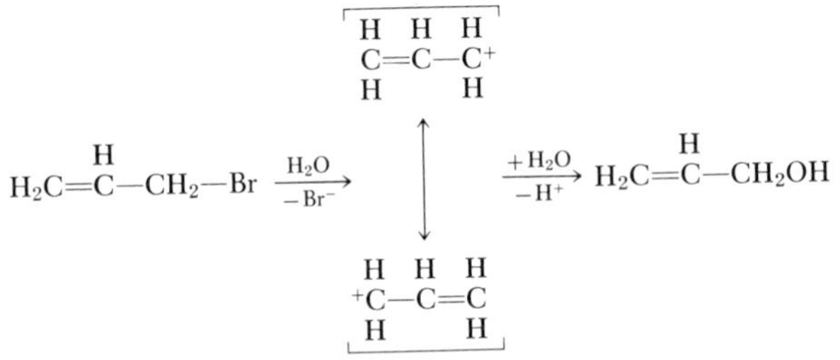

The carbonium ion formed here is a particularly stable one because of resonance stabilization of the ion, that is, the delocalization of the positive charge (see Figure 3.26 for the resonance of the corresponding negative ion).

The second type of hydrolysis occurs with alkyl halides such as the primary alkyl bromides which have high-energy carbonium ions (p. 89). Compounds of this type show little tendency to ionize in pure water. However, they react quite rapidly with hydroxide ions by a concerted process in which bond making (C—O bond formation) and bond breaking (C—Cl bond rupture) occur simultaneously.

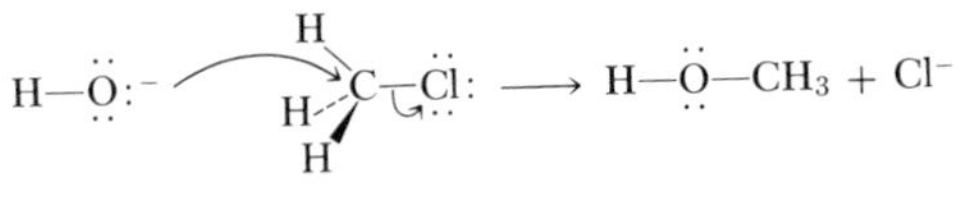

Alcohol formation also occurs with the nucleophile water, but the rate is considerably lower. Hydroxide ion is a stronger base than is water, and it is therefore a better nucleophile. Strictly speaking, basicity is a measure of the ability to abstract protons—for example, in acid-base reactions,

$$H-\ddot{O}:^- \longrightarrow H-\ddot{O}-N=\ddot{O}: \rightleftharpoons H_2O + {}^-:\ddot{O}-N=\ddot{O}:$$

and nucleophilicity is a measure of the rate of attack on a carbon atom (in substitution reactions). However, as long as we are dealing with the same attacking atom in measuring basicity and nucleophilicity (oxygen in the present case), then the stronger the base, the stronger the nucleophile; an increase in the electron density on oxygen increases both the basicity and the nucleophilicity. This parallelism breaks down when different atoms are being compared; for instance, SH^- is a better nucleophile than is OH^-, but it is a weaker base.

Many nucleophiles and inorganic ions other than hydroxide ion react with the alkyl halides:

$$K^+I^- + CH_3Cl \longrightarrow CH_3I + K^+Cl^-$$

$$Na^+SH^- + CH_2CH_2Cl \longrightarrow CH_2CH_2SH + Na^+Cl^-$$
Ethanethiol

$$Na^+:C\equiv N:^- + CH_3CH_2Br \longrightarrow CH_3CH_2C\equiv N: + Na^+Br^-$$
Ethyl cyanide

$$:NH_3 + CH_3CH_2CH_2Br \longrightarrow CH_3CH_2CH_2NH_3^+Br^-$$
$$\xrightarrow[\text{NaOH}]{} CH_3CH_2CH_2\ddot{N}H_2 + Na^+Br^- + H_2O$$
Propylamine

Reactions of this type are very important for the synthesis of organic compounds; they are called *nucleophilic substitution reactions* in view of the involvement of nucleophiles and the substitution nature of the reaction.

Nucleophilic substitutions are common in biological systems. As examples, methyl transfers (pp. 134, 153) and detoxification in the body of benzyl chloride can be cited. In the latter case, substitution occurs by glutathione,

$$
\begin{array}{cc}
\overset{\displaystyle CO_2H}{\underset{\displaystyle |}{}} & \overset{\displaystyle CH_2SH}{\underset{\displaystyle |}{}} \\
H_2NCCH_2CH_2CONHCCONHCH_2CO_2H \\
\underset{\displaystyle H}{|}
\end{array}
$$

(symbolized RSH), in an enzyme-catalyzed reaction to give the bound compound *a* (p. 92):

$$\langle\!\!\!\bigcirc\!\!\!\rangle-CH_2-Cl + RSH \longrightarrow \langle\!\!\!\bigcirc\!\!\!\rangle-CH_2-S-R$$
a

The reaction of a strong base with an alkyl halide leads to concurrent substitution and elimination reactions (see section on alkenes, Chapter 3).

$$
\begin{array}{ccc}
H & H & H \\
| & | & | \\
HC & -C & -C-Cl + Na^+OH^- \\
| & | & | \\
H & H & H
\end{array}
\longrightarrow
\begin{cases}
\longrightarrow H_3C-CH=CH_2 + Na^+Cl^- + H_2O \\
\longrightarrow H_3C-CH_2-CH_2-OH + Na^+Cl^-
\end{cases}
$$

A number of variables influence the ratio of elimination to substitution, but one of the most important is the degree of branching of the alkyl halides. Highly branched alkyl halides such as tertiary butyl chloride produce largely the alkene (2-methyl-2-propene, in this case), whereas the primary alkyl halides yield largely the alcohol.

Alcohols are also formed from Grignard reagents (p. 108), and from the addition of water to a double bond. Besides the mechanistic details for the latter reaction as presented in Chapter 3, it is known that the addition occurs in a *trans* sense. That is, in the acid-catalyzed addition to 1,2-dimethylcyclopentene, a proton attacks one face of the π electrons and a water molecule attacks the other face, to give *trans*-1,2-dimethylcyclopentanol (the *cis* isomer is not formed):

1,2-Dimethyl-
cyclopentene   *trans*   *cis*

Trans addition similarly occurs in biological hydrations catalyzed by the enzymes fumarase and aconitase; further subtleties of this reaction will be covered in Chapter 6. Lastly, the fermentation of glucose catalyzed by yeast enzymes to give potable ethanol should be mentioned as a biological source of alcohols.

$$C_6H_{12}O_6 \longrightarrow 2\,CO_2 + 2\,CH_3CH_2OH$$

REACTIONS OF THE ALCOHOLS   Because of the structural relationship of the alcohols to water and the inertness of hydrocarbon chains, most of the reactions of the alcohols are similar to those of water. Alcohols react with alkali metals to give salts, for example: $2\,ROH + 2\,Na \longrightarrow 2\,RO^-Na^+ + H_2$. Salts are named by substituting the suffix *-oxide* for the alcohol suffix *-anol* and adding the name of the positive ion. Examples are sodium methoxide, $CH_3O^-Na^+$, and sodium ethoxide, $CH_3CH_2O^-Na^+$ (collectively, these salts are referred to as *alkoxides*).

The alcohols are about as weakly acidic as water ($ROH + H_2O \rightleftharpoons H_3O^+ + RO^-$; $K = 10^{-13}$ to 10^{-18}), whereas the phenols are considerably more acidic:

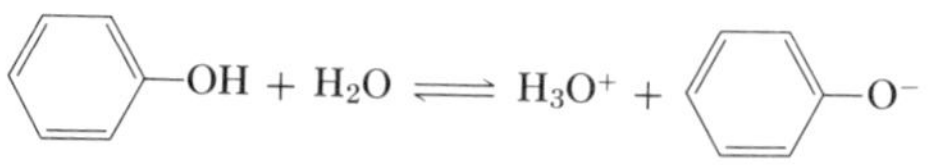

$$K = 10^{-10}$$

Because of this higher acidity, phenols will dissolve in water solutions of sodium hydroxide, whereas the high-molecular-weight alcohols are not dissolved by aqueous bases.

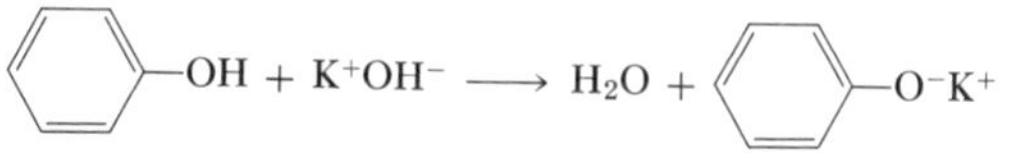

Mixtures of phenols and alcohols can be readily separated, therefore, with a dilute solution of sodium hydroxide. The phenol enters the aqueous phase as the sodium salt whereas the alcohol remains as a separate layer that can be removed. The water layer is then acidified to regenerate the phenol.

The higher acidity of phenol is attributed to the resonance of the phenoxide ion:

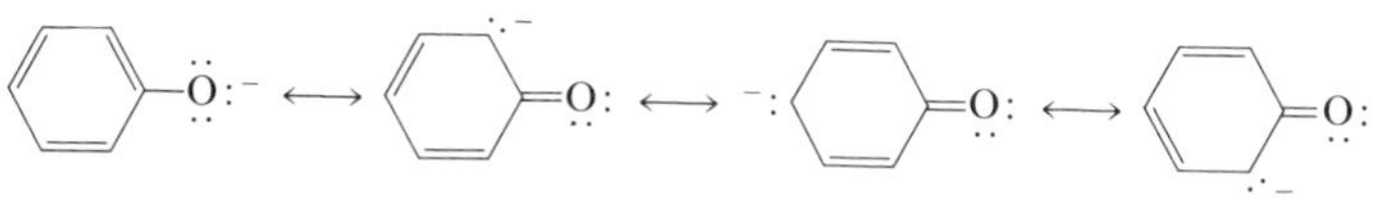

The negative charge is not localized on the oxygen atom, as it is in the alkoxide ions, R—O⁻, but it is delocalized over the aromatic ring. This results in a low electron density on oxygen, a low basicity of the negative ion, and a high phenoxide concentration at equilibrium. A more detailed explanation of the relationship of acidity to the structure of acids will be given in a later section on the acid strengths of the carboxylic acids.

Oxidation   Primary° alcohols are oxidized by potassium dichromate to give compounds called aldehydes:

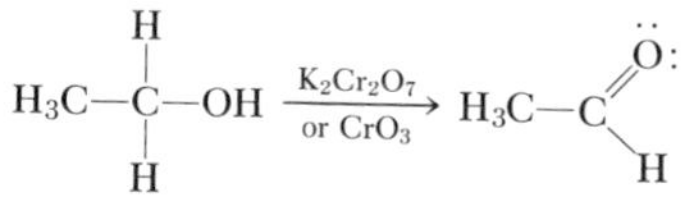

Acetaldehyde

and secondary alcohols yield compounds of a similar type called ketones:

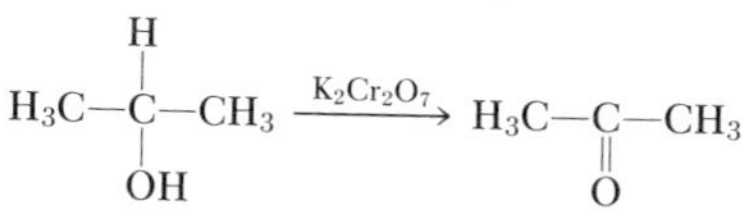

Acetone

° See p. 89 for the meaning of the terms *primary*, *secondary*, and *tertiary*.

Tertiary alcohols, in contrast, are stable to these oxidizing agents under normal conditions.

$$H_3C-\underset{\underset{CH_3}{|}}{\overset{\overset{CH_3}{|}}{C}}-OH \xrightarrow{K_2Cr_2O_7} \text{no oxidation}$$

The chemical oxidations appear to proceed through ester intermediates that yield the carbonyl group by a version of the elimination reaction (discussed in the section on alkenes in Chapter 3):

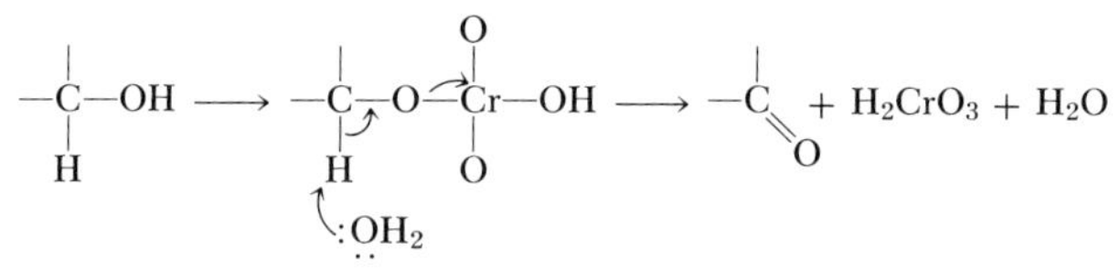

$$-\underset{\underset{H}{|}}{\overset{|}{C}}-OH \longrightarrow -\underset{\underset{H}{|}}{\overset{|}{C}}-O-\underset{\underset{O}{||}}{\overset{\overset{O}{||}}{Cr}}-OH \longrightarrow -\underset{\underset{O}{\diagdown}}{\overset{|}{C}} + H_2CrO_3 + H_2O$$

Enzymes called dehydrogenases catalyze the conversion of alcohols to ketones and aldehydes, but the mechanism is quite different from the chemical one outlined above. The coenzyme° DPN^+ is required and it appears that the transfer of a hydride ion $(H:^-)$ is involved in this further version of the elimination reaction (the enzyme contains zinc ions as part of its structure; a possible role for the zinc is shown here):

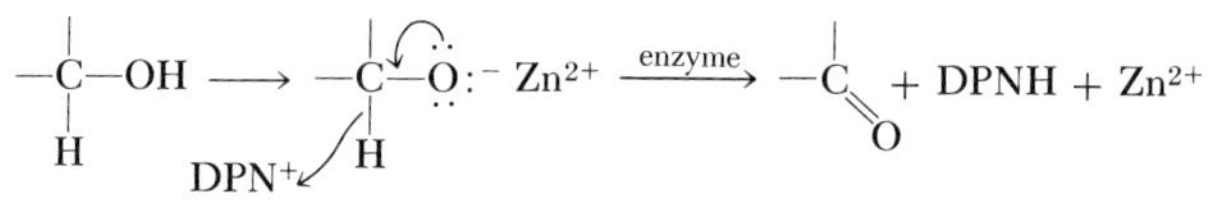

$$-\underset{\underset{H}{|}}{\overset{|}{C}}-OH \longrightarrow -\underset{\underset{DPN^+}{}}{\overset{|}{C}}-\overset{..}{O}:^- \; Zn^{2+} \xrightarrow{enzyme} -\underset{\underset{O}{\diagdown}}{\overset{|}{C}} + DPNH + Zn^{2+}$$

The oxidation of primary alcohols with an excess of $K_2Cr_2O_7$ or with $KMnO_4$ proceeds to a further stage to give the corresponding carboxylic acid:

$$R-CH_2-CH_2-OH \xrightarrow{KMnO_4} R-CH_2-\overset{\overset{O}{||}}{C}-OH$$

These oxidation reactions are utilized in the degradation of complex compounds. The formation of aldehyde or acid groups on oxidation would suggest, for example, that the compound contained primary alcohol groups, whereas the formation of ketone functions would indicate the presence of secondary alcohol groups.

The chemistry of the aldehydes, ketones, and acids will be outlined in later sections of this chapter.

° Coenzymes are relatively small molecules that together with the enzyme catalyze specific reactions. DPN^+ (also called NAD) is a complex compound with the nucleotide structure (p. 179).

Esterification One of the most valuable of the alcohol reactions is the formation from inorganic and organic acids of compounds known as esters:

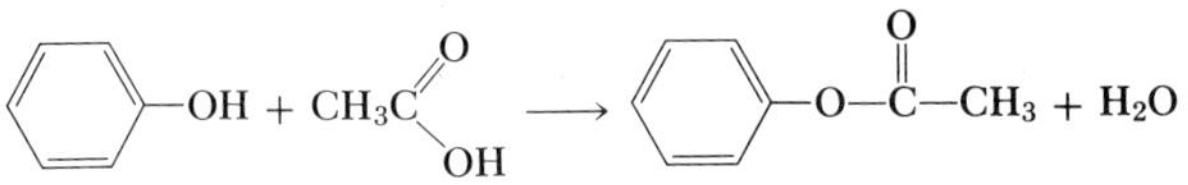

Phenyl acetate

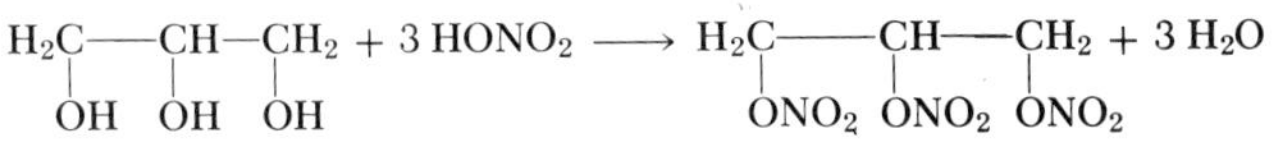

Glyceryl trinitrate
(nitroglycerine)

The esters are named as if they were salts (from the name of the organic radical and the name of the negative ion of the acid), although in reality they are covalently bonded compounds. The chemistry of these compounds will be discussed in greater detail in a later section of this chapter.

Ethers (ROR) are derivatives of water in which both of the hydrogen atoms are replaced by *alkyl* or *aryl*° groups. They are named in two different ways: either by the attachment of the names of the alkyl radicals to the generic name *ether*, or by a system in which they are treated as alkoxide derivatives of the hydrocarbons. Most of the *cyclic* ethers have common names, however. Examples are given in Figure 4.2.

The ethers are widely distributed in nature, and a few—vanillin, for example, which is the principle flavoring agent of vanilla (the structure is shown on p. 100)—are of commercial importance. The lower-molecular-weight ethers are used principally as solvents, although diethyl ether finds wide

° An aryl group is a radical derived from an aromatic hydrocarbon.

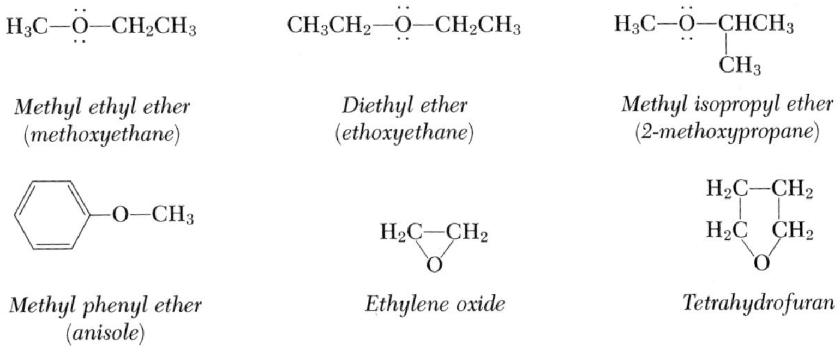

H_3C—$\overset{..}{\underset{..}{O}}$—$CH_2CH_3$ CH_3CH_2—$\overset{..}{\underset{..}{O}}$—$CH_2CH_3$ H_3C—$\overset{..}{\underset{..}{O}}$—$\underset{\underset{CH_3}{|}}{C}HCH_3$ *Figure 4.2*   **Common ethers.**

Methyl ethyl ether   Diethyl ether   Methyl isopropyl ether
(methoxyethane)   (ethoxyethane)   (2-methoxypropane)

Methyl phenyl ether   Ethylene oxide   Tetrahydrofuran
(anisole)

use in addition as an anesthetic.

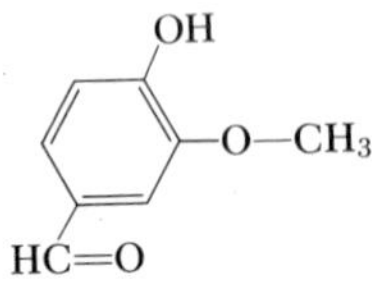

Vanillin

The ethers are conveniently prepared by the reaction of alkoxide ions with alkyl halides:

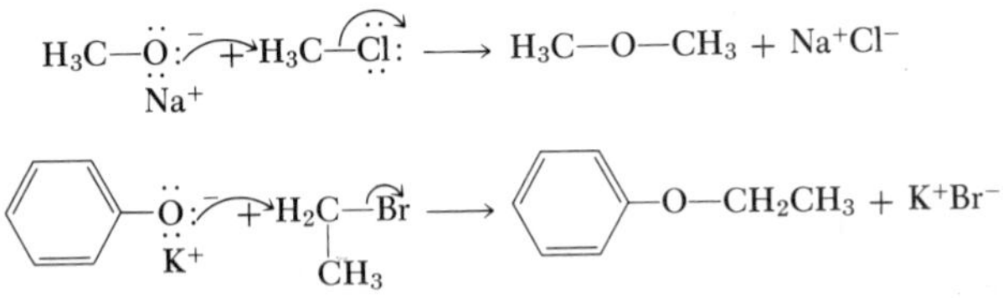

$$H_3C-\overset{..}{\underset{..}{O}}: \quad + H_3C-\overset{..}{\underset{..}{Cl}}: \longrightarrow H_3C-O-CH_3 + Na^+Cl^-$$
$$Na^+$$

This is a typical nucleophilic displacement reaction of an alkyl halide, and it is restricted essentially to primary and secondary alkyl halides (see section on the preparation of alcohols from alkyl halides).

As one might expect from their structures, pure ethers are not hydrogen-bonded, and as a result, their boiling points are similar to those of the related alkanes. The ethers are weak Lewis bases, however, and they do form hydrogen bonds with alcohols,

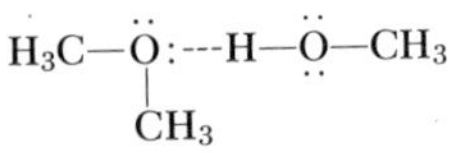

$$H_3C-\overset{..}{\underset{CH_3}{O}}: \text{---} H-\overset{..}{\underset{}{O}}-CH_3$$

salts with strong protic acids,

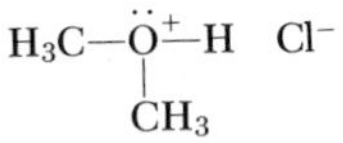

$$H_3C-\overset{+}{\underset{CH_3}{O}}-H \quad Cl^-$$

and salts with Lewis acids,

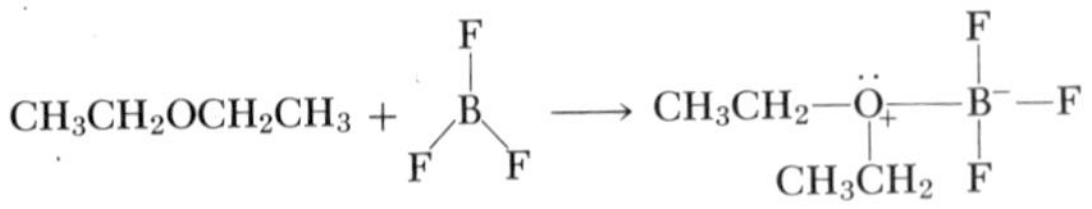

(BF$_3$ contains a boron atom with six electrons in its valence shell). Because of the basic properties of ethers, they may be cleaved by strong acids (protonation weakens the O—CH$_3$ bond, permitting a rapid displacement reaction):

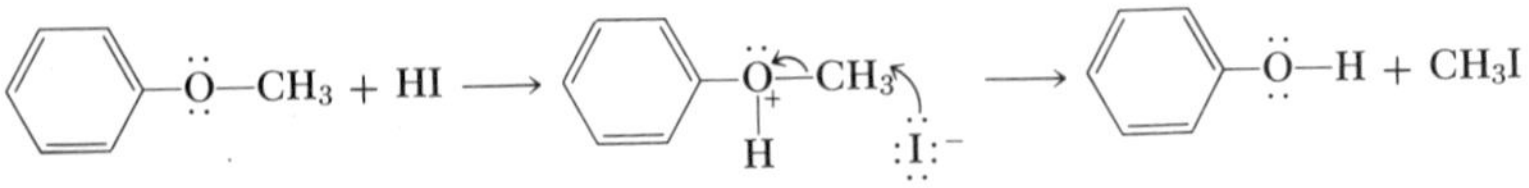

Since methoxy groups are common in compounds found in nature, the treatment of these compounds with hydrogen iodide and the isolation of the methyl iodide formed is very useful in studies aimed at determining their structures.

In contrast to this behavior with acids, ethers are very stable to bases, and unstable alcohols are often converted into ethers to protect the molecules from decomposition in certain reactions that would normally destroy the free alcohol (such as oxidation in basic solutions).

Aldehydes and *ketones* are alkyl and aryl derivatives of the simplest organic compound containing doubly bonded oxygen, formaldehyde

$$CH_2O \quad \text{or} \quad \overset{\ddot{O}}{\underset{}{H-C-H}}$$

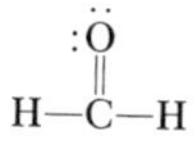

The aldehydes (RCHO) are monoalkyl derivatives and the ketones (RCOR) are dialkyl derivatives of formaldehyde. The aldehydes are named systematically by the substitution of the suffix *-al* for the suffix *-e* of the hydrocarbon molecule with the same chain length, and the ketones, by the substitution of the suffix *-one* for this ending (although in certain cases the *-e* is retained). For the structures of some simple aldehydes and ketones and their names (systematic and common), see Figure 4.3.

The aldehyde and ketone functional groups are common in natural products:

$CH_3(CH_2)_{12}CHO$

Tetradecanal

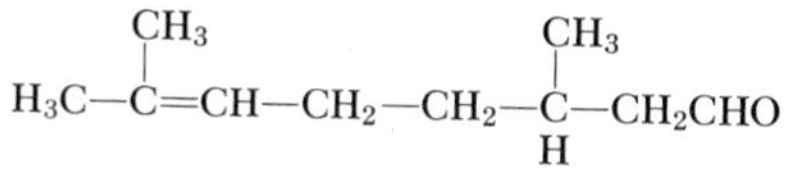

$$H_3C-\overset{CH_3}{\underset{H}{C}}=CH-CH_2-CH_2-\overset{CH_3}{\underset{H}{C}}-CH_2CHO$$

Citronellal

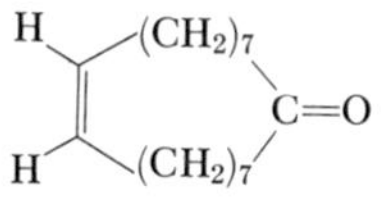

$$\begin{array}{c} H \diagdown \quad (CH_2)_7 \\ \qquad \qquad \diagup \\ \qquad \qquad C=O \\ H \diagup \quad (CH_2)_7 \diagup \end{array}$$

Civetone

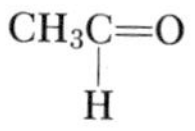

$$CH_3\overset{}{\underset{H}{C}}=O$$

Ethanal
(acetaldehyde)

$$H_3C-\overset{CH_3}{\underset{H}{C}}-CH_2-\overset{}{\underset{H}{C}}=O$$

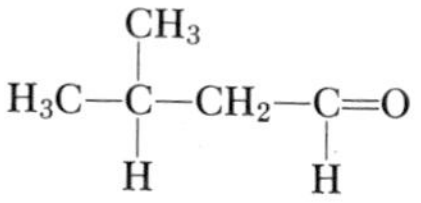

3-Methylbutanal
(3-methylbutyraldehyde)

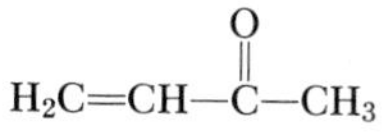

$$H_2C=CH-\overset{O}{\underset{}{C}}-CH_3$$

1-Butene-3-one

Figure 4.3 *Representative aldehydes and ketones.*

$$H_3C-\overset{O}{\underset{}{C}}-CH_3$$

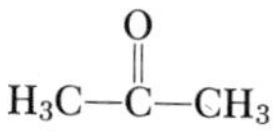

2-Propanone
(acetone)

1,3-Cyclopentanedione

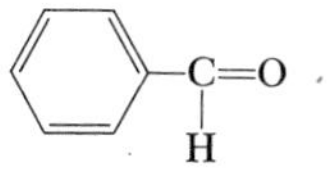

$$\overset{}{\underset{H}{C}}=O$$

Benzaldehyde

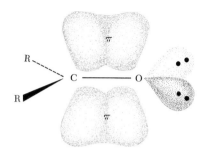

Figure 4.4 **Structure of the carbonyl group.**

Tetradecanal is required for light production in the luminous bacterium *Achromobacter fischeri*, citronellal is used by ants as a chemical alarm signal, and civetone is a perfume ingredient obtained from the scent glands of the civet cat. In addition, as will be shown in Chapter 6, most sugars contain an aldehyde group.

PREPARATION OF ALDEHYDES AND KETONES   The most general reaction for the preparation of aldehydes and ketones is the oxidation of alcohols with sodium dichromate or chromic oxide (see section on alcohols). The aromatic ketones are often prepared, in addition, by the *Friedel-Crafts reaction*, during which an aromatic compound reacts with a halogen derivative, with aluminum chloride as a catalyst:

$$R-\overset{\overset{\textstyle O}{\|}}{C}-Cl + AlCl_3 \longrightarrow R-\overset{\overset{\textstyle O}{\|}}{C}{}^+AlCl_4{}^-$$

This is another example of electrophilic substitution (see p. 87).

THE STRUCTURE OF ALDEHYDES AND KETONES   The doubly bonded carbon-oxygen group, called the *carbonyl group*, has a structure similar to that of the carbon–carbon double bond (Figure 4.4). The two R groups and the two unshared electron pairs on the oxygen atom are in one plane and the π electrons are perpendicular to that plane. The bond is a polar one because of the greater electronegativity of the oxygen atom; this is shown by means of the principal resonance contributors of the carbonyl group.

Tautomerism   Electrons on the α carbon (the carbon atom adjacent to the carbonyl group) can be delocalized over the carbonyl group:

Negative ions (called *anions*) of this type are far more stable than negative ions that are not resonance-stabilized, for example, the anion derived from methane

As a result, aldehydes and ketones are far more acidic than the hydrocarbons.

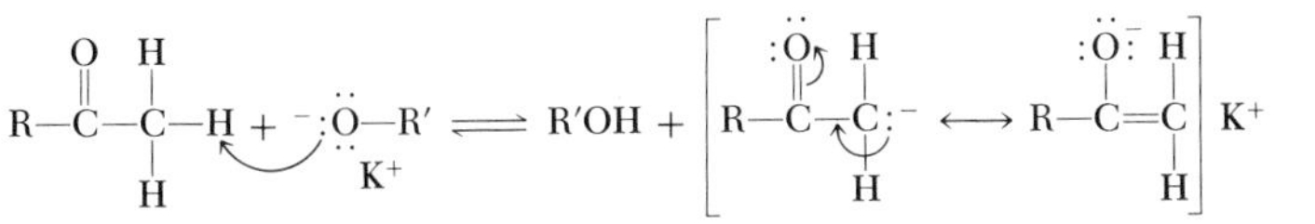

$$R-\overset{\overset{O}{\|}}{C}-\overset{\overset{H}{|}}{\underset{H}{C}}-H + \overset{..}{:}\overset{..}{O}-R' \underset{K^+}{\rightleftharpoons} R'OH + \left[R-\overset{\overset{:\overset{..}{O}}{\|}}{C}\curvearrowleft\overset{H}{\underset{H}{C}}:^- \longleftrightarrow R-\overset{\overset{:\overset{..}{O}^-}{|}}{C}=\overset{H}{\underset{H}{C}} \right] K^+$$

The ion formed in this way from aldehydes and ketones is called an *enolate ion.* The acidification of enolate ions yields the parent carbonyl compound and often, in addition, an isomeric compound.

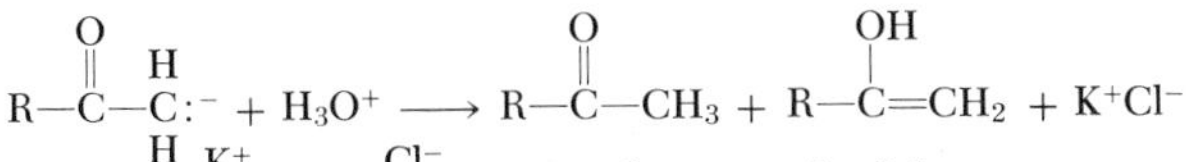

$$R-\overset{\overset{O}{\|}}{C}-\underset{\underset{K^+}{H}}{\overset{H}{C}}:^- + H_3O^+ \underset{Cl^-}{\longrightarrow} \underset{Keto\ form}{R-\overset{\overset{O}{\|}}{C}-CH_3} + \underset{Enol\ form}{R-\overset{\overset{OH}{|}}{C}=CH_2} + K^+Cl^-$$

The isomeric compound is called an *enol* (*en-* for the double bond and *-ol* for the hydroxy group). In a short time, the enol forms of most aldehydes and ketones change back into the normal carbonyl, or keto, forms; that is, the enol form is usually less stable than is the keto form.

The formation of enols may also be catalyzed by acids:

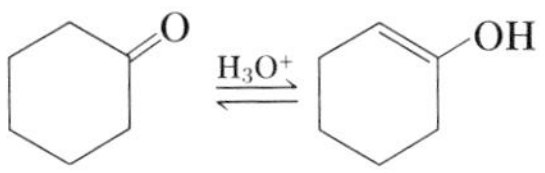

The general phenomenon of two isomeric species in equilibrium is called *tautomerism.* The equilibrium amount of the enol form in the tautomerism of simple aldehydes and ketones (acetone) is usually less than 1 percent at room temperature. The equilibrium amount of the enol form can approach 100 percent, however, in the case of diketones in which the enol form is stabilized by hydrogen-bonding (as in acetylacetone).

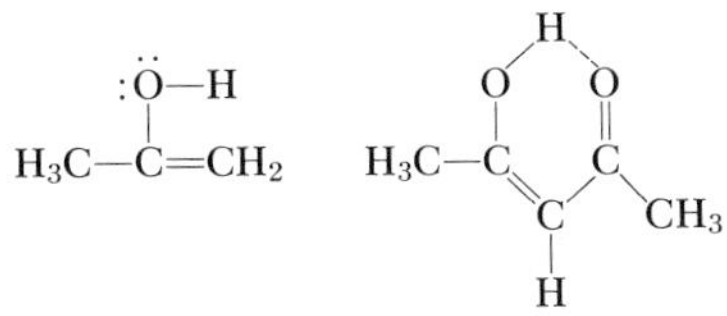

Enol of acetone   *Acetylacetone*

REACTIONS OF ALDEHYDES AND KETONES   *Reactions of the α-carbon atom*   Enols and enolate ions are highly reactive species that lead to ready reaction at the α positions. The base- and acid-catalyzed brominations of acetone are examples (Figure 4.5). In both types of bromination, the first step is a typical acid-base reaction, and the last step is a nucleophilic displacement on bromine of bromide ion.

Reactions of the carbonyl group: acetals and ketals   Compounds containing carbonyl groups react not only at the α carbon atom, but also at the carbonyl group; reactions of the latter type involve, typically, the addition

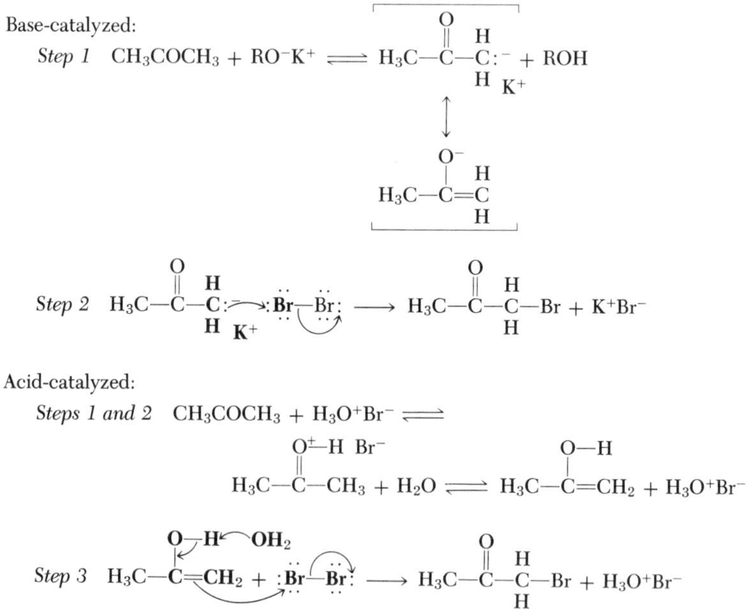

Base-catalyzed:

Step 1 $CH_3COCH_3 + RO^-K^+ \rightleftharpoons$

$$H_3C-\overset{\overset{\displaystyle O}{\|}}{C}-\overset{\overset{\displaystyle H}{|}}{\underset{\underset{\displaystyle H}{|}}{C}}\!:^- + ROH$$
K^+

$$H_3C-\overset{\overset{\displaystyle O^-}{|}}{C}=\overset{\overset{\displaystyle H}{|}}{\underset{\underset{\displaystyle H}{}}{C}}$$

Step 2 $H_3C-\overset{\overset{\displaystyle O}{\|}}{C}-\overset{\overset{\displaystyle H}{|}}{\underset{\underset{\displaystyle H}{|}}{C}}\!:\;\;:\!Br-Br: \longrightarrow H_3C-\overset{\overset{\displaystyle O}{\|}}{C}-\overset{\overset{\displaystyle H}{|}}{\underset{\underset{\displaystyle H}{|}}{C}}-Br + K^+Br^-$
K^+

Acid-catalyzed:

Steps 1 and 2 $CH_3COCH_3 + H_3O^+Br^- \rightleftharpoons$

$$H_3C-\overset{\overset{\displaystyle O^+-H}{\|}}{C}-CH_3 + H_2O \rightleftharpoons H_3C-\overset{\overset{\displaystyle O-H}{|}}{C}=CH_2 + H_3O^+Br^-$$

Step 3 $H_3C-\overset{\overset{\displaystyle O-H}{|}}{C}=CH_2 + :Br-Br: \longrightarrow H_3C-\overset{\overset{\displaystyle O}{\|}}{C}-\overset{\overset{\displaystyle H}{|}}{\underset{\underset{\displaystyle H}{|}}{C}}-Br + H_3O^+Br^-$

*Figure 4.5   **The base- and acid-catalyzed brominations of acetone.***

of some molecule across the multiple bond, just as the reactions of alkenes are characterized by addition across the carbon–carbon double bond. In water solutions of carbonyl compounds, for example, a low concentration of the *hydrate* (an addition compound of water) is formed:

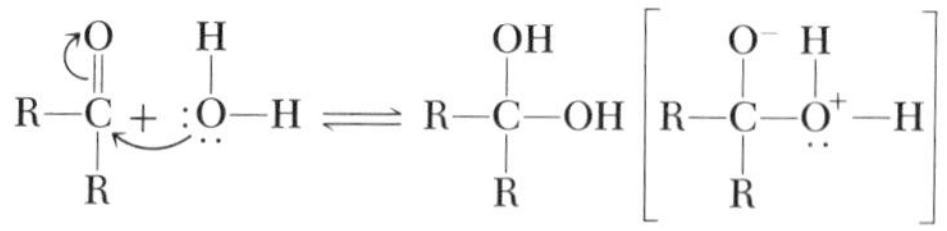

$$R-\overset{\overset{\displaystyle O}{\|}}{\underset{\underset{\displaystyle R}{|}}{C}} + :O-H \rightleftharpoons R-\overset{\overset{\displaystyle OH}{|}}{\underset{\underset{\displaystyle R}{|}}{C}}-OH \quad \left[R-\overset{\overset{\displaystyle O^-}{|}}{\underset{\underset{\displaystyle R}{|}}{C}}-\overset{\overset{\displaystyle H}{|}}{O^+}-H \right]$$
X

The first product formed during the addition reaction is actually species X; proton transfers are rapid, however, and often intermediates that can be converted to products solely by proton transfers are omitted from chemical equations. The related hydration of carbon dioxide catalyzed by the enzyme carbonic anhydrase was mentioned in Chapter 2. This reaction is utilized by the body to maintain the desired level of acidity in blood, in urine, and in other fluids. Such regulation of pH is possible because CO_2 is not an acid, whereas the hydrated form, carbonic acid (H_2CO_3), is an acid. Thus hydration in this case leads ultimately to the formation of hydronium ions.

Addition reactions of the carbonyl group differ from the addition reactions of the olefins because of the polarization of the carbonyl group, which leads to a partial positive charge on the carbon atom (Figure 4.6). As a result,

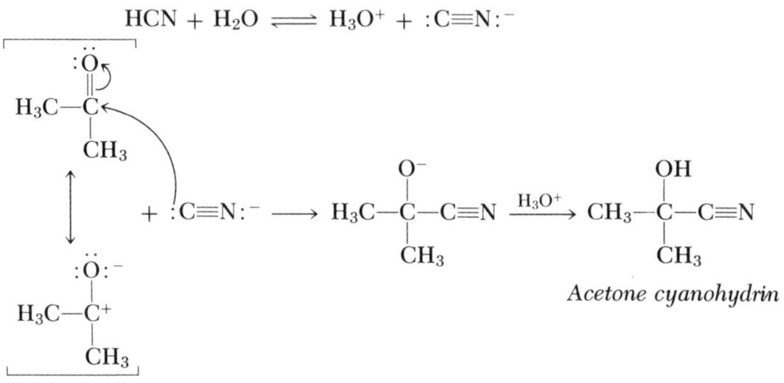

$$HCN + H_2O \rightleftharpoons H_3O^+ + :C\equiv N:^-$$

$$H_3C-\overset{\displaystyle :\overset{..}{O}:}{\underset{\displaystyle CH_3}{C}} \quad + \; :C\equiv N:^- \longrightarrow H_3C-\overset{\displaystyle O^-}{\underset{\displaystyle CH_3}{C}}-C\equiv N \xrightarrow{\text{H}_3\text{O}^+} CH_3-\overset{\displaystyle OH}{\underset{\displaystyle CH_3}{C}}-C\equiv N$$

$$H_3C-\overset{\displaystyle :\overset{..}{O}:^-}{\underset{\displaystyle CH_3}{C^+}}$$

Acetone cyanohydrin

Figure 4.6 **A nucleophilic addition reaction of acetone.**

species bearing unshared electrons (nucleophiles) can attack the bond directly (Figure 4.6), whereas in general, the alkenes do not react directly with nucleophiles. The addition reactions of carbonyl compounds may thus be classified as *nucleophilic addition reactions*.

We have now covered the major classes or organic reaction mechanisms; for reference purposes, a concise summary of these mechanisms is given in Table 4.2.

Table 4.2 **A summary of the major classes of organic reactions**

REACTIONS	CATALYZED BY OR INVOLVING	TYPES OF COMPOUNDS UNDERGOING THESE REACTIONS	PAGES IN TEXT COVERING THIS MATERIAL
I. *Free-radical reactions*	—	Alkanes, and alkane groups in other compounds	65–66
II. *Ionic reactions*			
(a) *Substitution reactions*			
(1) *Electrophilic*	Acids	Aromatic compounds	87–92
(2) *Nucleophilic*	Bases	Alkyl halides and related derivatives	96–98, 151–153, 172–174
(b) *Addition reactions*			
(1) *Electrophilic*	Acids or electrophiles	Alkenes	72–73, 88–89
(2) *Nucleophilic*	Bases or nucleophiles	Carbonyl compounds (and compounds containing C=N and C≡N groups)	105–106, 109
(c) *Elimination reactions*			
(1) *Electrophilic*	Acids	Alcohols	70–71
(2) *Nucleophilic*	Bases	Alkyl halides and related derivatives	71–72

107

Alcohol molecules can also add to the carbonyl group—as may be expected from the similarities between alcohols and water. The adduct in this case is called a *hemiacetal*.

$$R-\underset{\underset{\displaystyle O}{\|}}{C}-H + R'OH \rightleftharpoons R-\underset{\underset{\displaystyle OR'}{|}}{\overset{\overset{\displaystyle OH}{|}}{C}}-H$$

If a solution of the hemiacetal in an excess of alcohol is treated with acids, another alcohol molecule participates, and a double ether called an *acetal* is formed:

$$CH_3CHO + CH_3OH \rightleftharpoons H_3C-\underset{\underset{\displaystyle O-CH_3}{|}}{\overset{\overset{\displaystyle OH}{|}}{C}}-H \xrightarrow{+H_3O^+} H_2O + H_3C-\underset{\underset{\displaystyle O-CH_3}{|}}{\overset{\overset{\displaystyle H-O^+-H}{|}}{C}}-H \longrightarrow$$

$$H_2O + \left[H_3C-\underset{\underset{\displaystyle :\ddot{O}-CH_3}{|}}{\overset{\overset{\displaystyle +}{|}}{C}}-H \longleftrightarrow H_3C-\underset{\underset{\displaystyle +\ddot{O}-CH_3}{\|}}{C}-H \right] \xrightarrow[-H^+]{CH_3OH} H_3C-\underset{\underset{\displaystyle OCH_3}{|}}{\overset{\overset{\displaystyle OCH_3}{|}}{C}}-H$$

Acetaldehyde dimethyl acetal

This reaction proceeds through a carbonium ion, which is a particularly stable one because of the resonance interaction of the unshared electron pair on oxygen. The attack of an alcohol molecule acting as a nucleophile on this ion leads to the acetal. The corresponding product from a ketone is called a *ketal*.

Ketals and acetals do not have the chemical properties of carbonyl compounds; instead, they behave as typical ethers. They are stable to bases, but they are cleaved by acids, and in aqueous acid solution they are readily reconverted into the corresponding carbonyl compounds.

Amines (RNH_2) and thiols (RSH) also add to carbonyl groups to form compounds related to the hemiacetals and ketals. Examples will be given in the section on sugars in Chapter 6.

The reaction of Grignard reagents with carbonyl compounds   Grignard reagents are organomagnesium compounds prepared by the action of metallic magnesium on the halogen derivatives of the hydrocarbons.

$$CH_3I + Mg \longrightarrow H_3C-Mg-I$$

$$CH_3CHBrCH_3 + Mg \longrightarrow H_3C-\underset{\underset{\displaystyle CH_3}{|}}{\overset{\overset{\displaystyle H}{|}}{C}}-Mg-Br$$

The carbon-metal bond is partially ionic, $R-Mg-X \longleftrightarrow R:^{-+}Mg-X$,

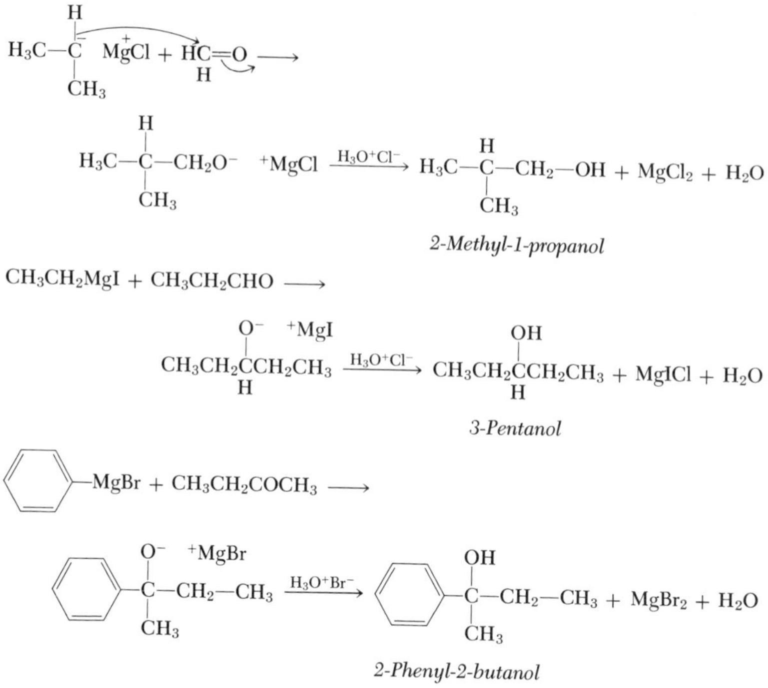

$$H_3C-\overset{\overset{\displaystyle H}{|}}{\underset{\underset{\displaystyle CH_3}{|}}{C}}\ \overset{+}{MgCl} + H\overset{}{C}=O \longrightarrow$$

$$H_3C-\overset{\overset{\displaystyle H}{|}}{\underset{\underset{\displaystyle CH_3}{|}}{C}}-CH_2O^- \ ^+MgCl \xrightarrow{H_3O^+Cl^-} H_3C-\overset{\overset{\displaystyle H}{|}}{\underset{\underset{\displaystyle CH_3}{|}}{C}}-CH_2-OH + MgCl_2 + H_2O$$

2-Methyl-1-propanol

$$CH_3CH_2MgI + CH_3CH_2CHO \longrightarrow$$

$$CH_3CH_2\overset{\overset{\displaystyle O^-\ ^+MgI}{|}}{\underset{\underset{\displaystyle H}{|}}{C}}CH_2CH_3 \xrightarrow{H_3O^+Cl^-} CH_3CH_2\overset{\overset{\displaystyle OH}{|}}{\underset{\underset{\displaystyle H}{|}}{C}}CH_2CH_3 + MgICl + H_2O$$

3-Pentanol

$$\text{⟨⟩}-MgBr + CH_3CH_2COCH_3 \longrightarrow$$

$$\text{⟨⟩}-\overset{\overset{\displaystyle O^-\ ^+MgBr}{|}}{\underset{\underset{\displaystyle CH_3}{|}}{C}}-CH_2-CH_3 \xrightarrow{H_3O^+Br^-} \text{⟨⟩}-\overset{\overset{\displaystyle OH}{|}}{\underset{\underset{\displaystyle CH_3}{|}}{C}}-CH_2-CH_3 + MgBr_2 + H_2O$$

2-Phenyl-2-butanol

Figure 4.7 *The reaction of Grignard reagents with aldehydes and ketones.*

and the carbon atom bonded to magnesium has some of the properties of a negative ion.

The reaction of Grignard reagents with carbonyl compounds is one of the best methods available for the synthesis of alcohols. The reaction involves addition of the Grignard reagent to the C=O double bond (Figure 4.7). A salt of the alcohol is formed as an intermediate in the reaction, but this derivative is readily hydrolyzed by dilute acids to yield the free alcohol. Primary, secondary, and tertiary alcohols may be prepared by this method (Figure 4.7).

Derivatives of aldehydes and ketones   In many nucleophilic addition reactions of carbonyl compounds, the intermediate addition compound cannot be isolated since the adduct readily loses a molecule of water. Reactions of this type involving hydrazine, hydroxylamine, phenylhydrazine, and related compounds are important since the derivatives ultimately formed (Figure 4.8) are usually crystalline, whereas common aldehydes and ketones are liquids. The melting point of a derivative of this type may serve to identify a carbonyl compound isolated from some reaction mixture or natural source since the melting points of the derivatives of most common carbonyl compounds are on record. The compound 2,4-dinitrophenylhydrazine is especially important in this respect in that its derivatives are readily obtained in the crystalline state, and they have high melting points.

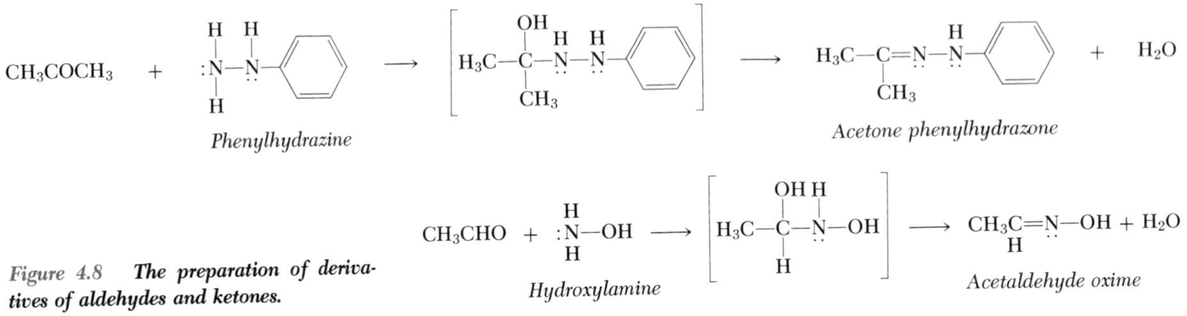

Figure 4.8 **The preparation of derivatives of aldehydes and ketones.**

Reduction of aldehydes and ketones Most carbonyl compounds are reduced to the corresponding alcohols by hydrogen in the presence of platinum and nickel catalysts, and also by lithium aluminum hydride:

$$\underset{\text{3-Pentanone}}{CH_3CH_2\overset{\displaystyle O}{\overset{\|}{C}}CH_2CH_3} \xrightarrow[\text{or } H_2 + \text{Pt or Ni}]{\text{LiAlH}_4} \underset{\text{3-Pentanol}}{CH_3CH_2\overset{\displaystyle OH}{\overset{|}{\underset{\underset{\displaystyle H}{|}}{C}}}CH_2CH_3}$$

The closest biological analogy to this reaction is the DPNH reduction of ketones to alcohols (both DPNH and $LiAlH_4$ are sources of $H:^-$). The DPNH reaction is reversible (see p. 100), and depending on the relative amounts of DPNH or DPN^+, one can force either the reduction of ketones or the oxidation of alcohols.

Reactions involving both the carbonyl group and the α carbon The anion formed by the loss of an α hydrogen of a ketone or aldehyde is a nucleophile, and it can add to the carbonyl group of another molecule in the normal way:

$$CH_3CHO + OH^- \rightleftharpoons {}^-{:}CH_2CHO + H_2O$$

$$H_3C-\overset{\displaystyle O}{\overset{\|}{C}}-H + {}^-{:}\overset{\displaystyle H}{\underset{\displaystyle H}{C}}-CHO \longrightarrow$$

$$H_3C-\overset{\displaystyle O^-}{\overset{|}{\underset{\underset{\displaystyle H}{|}}{C}}}-CH_2-CHO \xrightarrow[H_2O]{} \underset{\text{Aldol}}{H_3C-\overset{\displaystyle OH}{\overset{|}{\underset{\underset{\displaystyle H}{|}}{C}}}-CH_2-CHO} + OH^-$$

This reaction is known as the *aldol condensation*. The first step is the formation of an enolate ion, and the second is the addition of the ion to the carbonyl group of a second molecule. The products of the aldol condensation are β-hydroxy aldehydes or ketones. If these compounds are treated with tiny amounts of acids, which act as catalysts, an elimination reaction occurs to yield an α, β-unsaturated derivative.

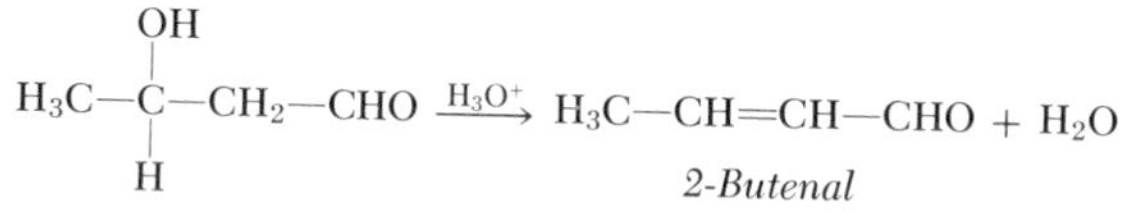

$$\underset{\underset{H}{|}}{\overset{\overset{OH}{|}}{H_3C-C}}-CH_2-CHO \xrightarrow{H_3O^+} H_3C-CH=CH-CHO + H_2O$$

$$\textit{2-Butenal}$$

The driving force for the elimination reaction is the formation of a set of double bonds which are resonance-stabilized:

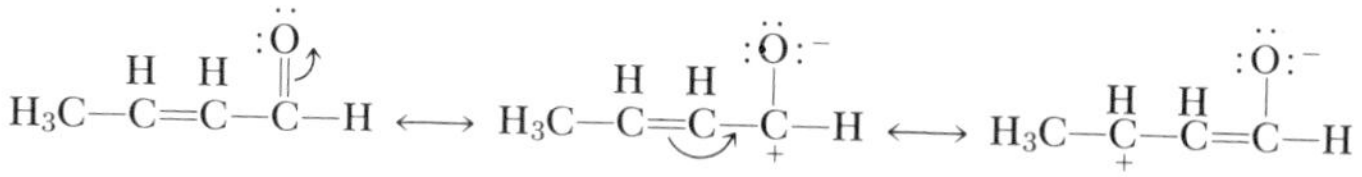

$$\underset{H_3C-C=C-C-H}{\overset{H\quad H\quad :\overset{..}{O}}{}} \longleftrightarrow \underset{H_3C-C=C-C-H}{\overset{H\quad H\quad :\overset{..}{O}:^-}{}} \longleftrightarrow \underset{H_3C-C-C=C-H}{\overset{H\quad H\quad :\overset{..}{O}:^-}{}}$$

Double bonds joined directly to one another are called *conjugated* double bonds. Because of the resonance-stabilization of conjugated systems, compounds with nonconjugated double bonds can usually be isomerized into the conjugated forms through the catalytic action of acids or bases:

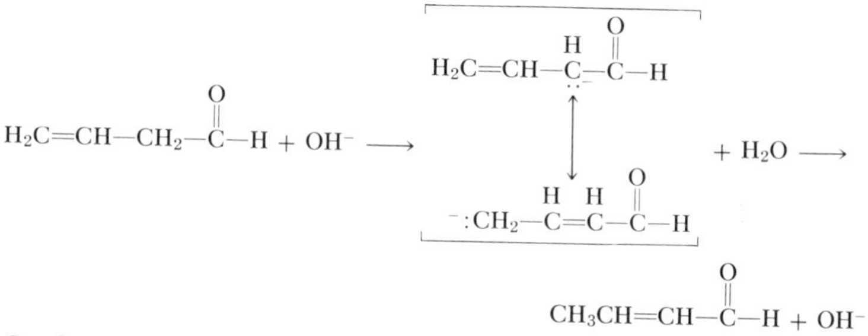

$$H_2C=CH-CH_2-\overset{\overset{O}{\|}}{C}-H + OH^- \longrightarrow$$

$$H_2C=CH-\overset{H}{\underset{..}{C}}-\overset{\overset{O}{\|}}{C}-H$$

$$^-:CH_2-\overset{H\quad H}{C=C}-\overset{\overset{O}{\|}}{C}-H$$

$$+ H_2O \longrightarrow$$

$$CH_3CH=CH-\overset{\overset{O}{\|}}{C}-H + OH^-$$

Condensations of the aldol type are well known in biological systems. As an example, the condensation of dihydroxyacetone phosphate (*a*) and 3-phosphoglyceraldehyde (*b*), catalyzed by the enzyme aldolase, gives fructose-1,6-diphosphate (*c*):

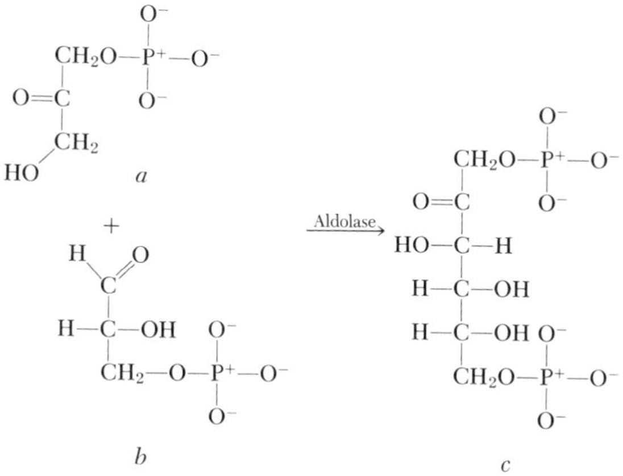

$$\begin{array}{c} \overset{O^-}{|} \\ CH_2O-P^+-O^- \\ | \\ O=C \quad O^- \\ | \\ CH_2 \\ HO \end{array}$$

$$a$$

$$+$$

$$\begin{array}{c} H \quad O \\ \diagdown C \diagup \\ | \\ H-C-OH \quad O^- \\ | \quad | \\ CH_2-O-P^+-O^- \\ | \\ O^- \end{array}$$

$$b$$

$$\xrightarrow{\text{Aldolase}}$$

$$\begin{array}{c} O^- \\ | \\ CH_2O-P^+-O^- \\ | \\ O=C \quad O^- \\ | \\ HO-C-H \\ | \\ H-C-OH \\ | \\ H-C-OH \quad O^- \\ | \quad | \\ CH_2O-P^+-O^- \\ | \\ O^- \end{array}$$

$$c$$

The formula of a monoalkyl phosphate is

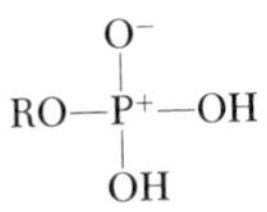

$$RO-\overset{\overset{\displaystyle O^-}{|}}{\underset{\underset{\displaystyle OH}{|}}{P^+}}-OH$$

but these compounds and phosphoric acid itself

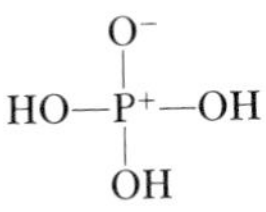

$$HO-\overset{\overset{\displaystyle O^-}{|}}{\underset{\underset{\displaystyle OH}{|}}{P^+}}-OH$$

are strong acids, and at the neutral pH values of body fluids the alkyl phosphates are essentially fully ionized as shown in *a*, *b*, and *c*. As we shall see in Chapter 6, it makes a difference whether we write the OH groups in *c* on the right side of the chain or on the left side. The reaction gives specifically the isomer shown.

Carboxylic acids (RCO_2H) are hydrocarbon derivatives in which one of the end carbon atoms is in a maximum state of oxidation.

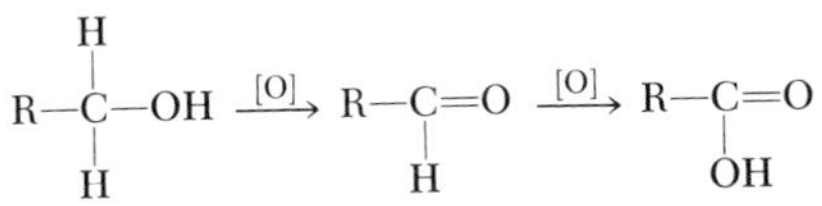

$$R-\overset{\overset{\displaystyle H}{|}}{\underset{\underset{\displaystyle H}{|}}{C}}-OH \xrightarrow{[O]} R-\overset{\overset{\displaystyle }{|}}{\underset{\underset{\displaystyle H}{|}}{C}}{=}O \xrightarrow{[O]} R-\overset{\overset{\displaystyle }{|}}{\underset{\underset{\displaystyle OH}{|}}{C}}{=}O$$

The $-CO_2H$ group is called a carboxyl group, and it is constructed of one carbonyl group and one hydroxyl group. The carbon atom is sp^2 hybridized, and therefore the two oxygen atoms, the carbon atom of the carbonyl group, and the α-carbon atom are in one plane (the OH bond can also be in this plane since there is free rotation about the bond joining it to the carbonyl group).

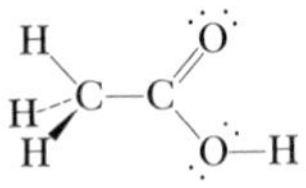

$$\overset{\displaystyle H}{\underset{\displaystyle H}{H{\diagdown}}}C-C\overset{\diagup\ddot{O}}{\diagdown\ddot{O}-H}$$

The carboxylic acids are easy to isolate from natural sources and quite a number of them have been known since the days of alchemy. Many of the acids were named in a nonsystematic way at that time, and a few of these names are still in use. Systematic names have been devised for the carboxylic acids, however; these are obtained by substituting the suffix *-oic* for the ending *-e* of the hydrocarbon of the same chain length, and adding the word "acid." In another system, the CO_2H group is treated as a substituent and the acid is named simply by adding the words "carboxylic acid" to the hydrocarbon name. Various common acids are listed in Figure 4.9 (the common, or trivial, names are given in parentheses).

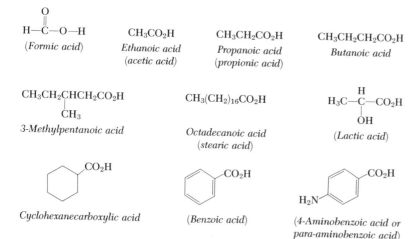

H—C—O—H
O
(*Formic acid*)

CH_3CO_2H
Ethanoic acid
(*acetic acid*)

$CH_3CH_2CO_2H$
Propanoic acid
(*propionic acid*)

$CH_3CH_2CH_2CO_2H$
Butanoic acid

$CH_3CH_2CHCH_2CO_2H$
        |
       CH_3
3-Methylpentanoic acid

$CH_3(CH_2)_{16}CO_2H$
Octadecanoic acid
(*stearic acid*)

H_3C—C—CO_2H
     |
    OH
    H
(*Lactic acid*)

Cyclohexanecarboxylic acid

(*Benzoic acid*)

(*4-Aminobenzoic acid or
para-aminobenzoic acid*)

Figure 4.9 **Common monocarboxylic acids.**

A large number of acids with two, three, or more carboxylic acid groups (di-, tri-, and polycarboxylic acids, respectively) occur in nature, where they are often responsible for the sourness of various unripened fruits; representative examples are given in Figure 4.10. These polycarboxylic acids are easily isolated from plant extracts; most of the aliphatic examples are also present in the human body, where they are involved as intermediates in the conversion of more complex compounds into carbon dioxide (the citric acid cycle).

CO_2H
|
CO_2H
Oxalic acid

H_2C
CO_2H
CO_2H
Malonic acid

H_2C—CO_2H
H_2C—CO_2H
Succinic acid

$(H_2C)_3$
CO_2H
CO_2H
Glutaric acid

Figure 4.10 **Common polycarboxylic acids.**

Fumaric acid

Maleic acid

HO—C—CO_2H
     |
    H
H_2C—CO_2H
Malic acid

H
|
HO—C—CO_2H
HO—C—CO_2H
|
H
Tartaric acid

H_2C—CO_2H
HO—C—CO_2H
H_2C—CO_2H
Citric acid

CO_2H
CO_2H
Phthalic acid

ORGANIC COMPOUNDS CONTAINING OXYGEN

ACID STRENGTH   The carboxylic acids are weak acids and their ionization constants may be readily calculated from the concentrations of the various species present at equilibrium:

$$R-\overset{\overset{\displaystyle O}{\parallel}}{C}-OH + H_2O \rightleftharpoons H_3O^+ + R-CO_2^-$$

$$K = \frac{[H_3O^+][RCO_2^-]}{[RCO_2H]}$$

K ranges from 10^{-1} to 10^{-6}, depending on R

The ionization constants are a measure of acid strength; the larger the constant, the stronger the acid.

The carboxylic acids, although they are weak, are far more acidic than are the alcohols:

$$R-O-H + H_2O \rightleftharpoons H_3O^+ + R-O^-$$

The greater acidity of the carboxylic acids is due largely to resonance-stabilization of the carboxylate ion:

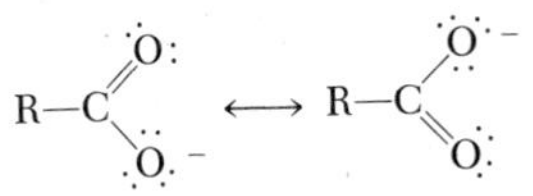

In the carboxylate ion, the negative charge is delocalized over two oxygen atoms, whereas in the alkoxide ion, the charge resides on a single oxygen atom. There is a higher electron density on oxygen in the alkoxide ion and therefore it is a stronger base. This relationship of electron density to base strength is also seen in a series of alkyl-substituted amines; alkyl groups are electron-releasing groups and the base strength is a function of the number of alkyl groups attached to nitrogen.

$$\text{Base strength:} \quad H_3C-\overset{\overset{\displaystyle H}{|}}{N}-CH_3 > H_3C-\overset{\overset{\displaystyle H}{|}}{N}H > H\overset{\overset{\displaystyle H}{|}}{N}H$$

Continuing with the explanation for the low acidity of alcohols relative to the carboxylic acids, since RO$^-$ is a stronger base than is RCO$_2^-$, it reacts faster with hydronium ions. This results in a lower concentration of H$_3$O$^+$ and a higher concentration of the neutral species at equilibrium (and a lower value of the ionization constant). That is, $K < K'$ for the following equilibria:

$$ROH + H_2O \underset{K}{\rightleftharpoons} H_3O^+ + RO^-$$

$$RCO_2H + H_2O \underset{K'}{\rightleftharpoons} H_3O^+ + RCO_2^-$$

An alternative thermodynamic view is that a greater free-energy change (ΔG) accompanies the formation of the resonance-*stabilized* carboxylate ion (RCO_2^-) than takes place in the ionization of the alcohol. From the equation connecting free energy and equilibrium constants (p. 50), K' must then be larger than K, and thus RCO$_2$H is a stronger acid than is ROH.

Table 4.3 Ionization constants of carboxylic acids

A. Aliphatic acids

ACID	FORMULA	K	
Propionic acid	$CH_3CH_2CO_2H$	1.4×10^{-5}	
Acetic acid	CH_3CO_2H	1.8×10^{-5}	
Iodoacetic acid	ICH_2CO_2H	7.5×10^{-4}	Acid strength
Chloroacetic acid	$ClCH_2CO_2H$	1.6×10^{-3}	
Trichloroacetic acid	Cl_3CCO_2H	2.0×10^{-1}	

B. Substituted benzoic acids

SUBSTITUENT	K	SUBSTITUENT	K	SUBSTITUENT	K
p-NH_2	1.2×10^{-5}	m-NH_2	1.6×10^{-5}	o-NH_2	1×10^{-5}
p-CH_3	4.2×10^{-5}	m-CH_3	5.4×10^{-5}	o-CH_3	1.2×10^{-4}
p-H	6.3×10^{-5}	m-H	6.3×10^{-5}	o-H	6.3×10^{-5}
p-NO_2	3.8×10^{-4}	m-NO_2	3.2×10^{-4}	o-NO_2	6.7×10^{-3}

This interpretation of the relative acidities of the carboxylic acids and alcohols suggests that electron-releasing groups should decrease the acidity of the carboxylic acids and that electron-attracting groups should increase the acidity. This is precisely the effect that is found. The ionization constants for several series of acids illustrating this point are given in Table 4.3.

The acidities of the acids vary over a wide range. The aliphatic acids listed are all derivatives of acetic acid, and therefore the differences in acidity must be due to the substituents on the α-carbon atom. Alkyl groups are electron-releasing groups, as will be recalled from the accounts of the stability order of carbonium ions (tertiary > secondary > primary) and of the direction in which water and the acids add to double bonds (Chapter 3). Consistent with the direction of these effects and with the interpretation of acid strength given in our comparison of alcohols and carboxylic acids, the substitution of one methyl group for an α hydrogen in acetic acid results in a decrease of the ionization constant from 1.8×10^{-5} to 1.4×10^{-5} (Table 4.3).

The halogens are electron-attracting substituents, on the other hand, and we see from Table 4.3 that chloroacetic acid is about 90 times as strong as acetic acid and that trichloroacetic acid is about 125 times as strong as monochloroacetic acid. The position of iodoacetic acid on our scale indicates a

ORGANIC COMPOUNDS CONTAINING OXYGEN

parallelism between the electronegativity of the halogen (pp. 33–34) and its effect in increasing the strength of substituted acetic acids.

We must also consider the effect of substituents on the rate of the forward reaction in the ionization of carboxylic acids:

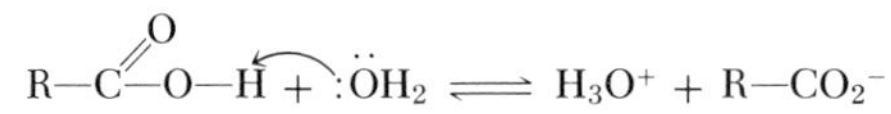

$$R-\overset{\displaystyle O}{\underset{}{C}}-O-H + :\ddot{O}H_2 \rightleftharpoons H_3O^+ + R-CO_2^-$$

Electron-attracting substituents should increase this rate since they should facilitate the separation of the developing positive charge on the proton and the negative charge on the oxygen. This effect happens to be in the same direction as the effect of electron-attracting substituents on the carboxylate ion; that is, both explanations lead to the conclusion that electron-attracting substituents should increase the strength of acids. The effect of substituents on neutral species is less, in general, than the effect on charged species, and for convenience, we often focus attention on only the negative ion in acid-base equilibria.

The inductive effect   The release and withdrawal of electrons by the substituent in the aliphatic acids is transmitted through the saturated α carbon atom of the acid; that is, through the σ bonds. The alteration of the electron density at the reaction site in this way is called the *inductive effect* and it is usually symbolized by an arrow indicating the direction of the electron displacement:

$$\overset{\displaystyle H}{\underset{\displaystyle H}{HC}} \longrightarrow CH_2 \longrightarrow CO_2H \qquad\qquad Cl \longleftarrow CH_2 \longleftarrow CO_2H$$

The inductive effect of a substituent (see illustration below) and the resonance effect (which is transmitted through the π electrons; see illustration b) are the

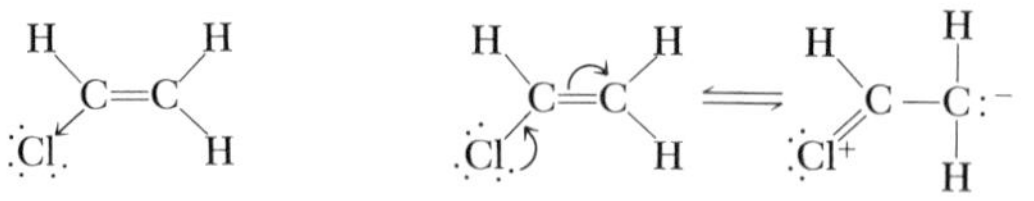

a b

principal ways in which the electronic structure of a compound may be altered by a substituent. In some cases, the effects are in opposite directions (as in the example of chloroethylene cited above), whereas in others, the effects are in the same direction (as in the methyl derivative of ethylene). In general, when the effects are in opposite directions, the resonance effect outweighs the inductive effect.

The effect of substituents on the acidity of benzoic acid is illustrated by the ionization constants in Table 4.3; the effects are similar to those discussed in the section on the aliphatic acids in that electron-attracting substituents lead to large values of K, whereas electron-releasing substituents lead to small values of K. Although the effect of substituents in the *para* and *meta* positions can be accounted for by the principles outlined, the effect of *ortho* substituents

(Table 4.3) is not so readily explained. *Ortho* substituents are very close to the carboxylic acid group and it is thought that such substituents crowd and distort the CO_2H group. This type of interaction is called the *steric effect;* another example is given in the next section.

It should be noted that those substituents which decrease the acid strength of benzoic acid activate the aromatic ring and lead to *ortho-para* electrophilic substitution, whereas those substituents that increase the acid strength deactivate the ring and lead to *meta* electrophilic substitution (see Chapter 3).

PREPARATION OF CARBOXYLIC ACIDS   The carboxylic acids may be prepared by the oxidation of primary alcohols and aldehydes,

$$RCH_2OH \text{ or } RCHO \xrightarrow{KMnO_4} RCO_2H$$

and also by the addition of Grignard reagents to carbon dioxide:

$$R:^-MgCl^+ + \ddot{:}\ddot{O}=C=\ddot{O}: \longrightarrow \left[R-\overset{\overset{\displaystyle O}{\|}}{C}-\ddot{O}:^-MgCl^+ \right] \xrightarrow{HCl} R-CO_2H + MgCl_2$$

This is another example of nucleophilic addition to a carbonyl group.

A third useful method involves the acid hydrolysis of an alkyl cyanide, prepared in turn by the displacement reaction of cyanide ion on an alkyl halide (bromide or iodide preferably):

$$R-Br + Na^+:C\equiv N:^- \longrightarrow Na^+Br^- + R-C\equiv N:$$

$$\xrightarrow[HCl]{H_2O} R-CO_2H + NH_4^+Cl^-$$

The biological synthesis of carboxylic acids usually involves the oxidation of an aldehyde. Energy is available from this oxidation, and although it is lost as heat in laboratory oxidations, living organisms are able to utilize the energy of oxidation. This is done by coupling the oxidation step (which has a negative free-energy change, $-\Delta G$; see Chapter 2) to the synthesis of ATP (adenosine triphosphate; formula appears on p. 156) from ADP (adenosine diphosphate) and inorganic phosphate (H_3PO_4) a process that has a positive free-energy change ($+\Delta G$). The ATP so formed can then be used elsewhere in the cell to provide energy for muscle contraction, nerve impulses, synthesis of vital cell constituents, and so on.

As an example of the energy coupling, the oxidation of 3-phosphoglyceraldehyde yields 3-phosphoglyceric acid, and the reaction has a ΔG^0 of about $-7,000$ cal/mole.

$$^-O-\overset{\overset{\displaystyle O^-}{|}}{\underset{\underset{\displaystyle O^-}{|}}{P^+}}-O-CH_2-\overset{\overset{\displaystyle H}{|}}{\underset{\underset{\displaystyle OH}{|}}{C}}-C\overset{\diagup O}{\diagdown_H} + H_2O - 2e \longrightarrow {}^-O-\overset{\overset{\displaystyle O^-}{|}}{\underset{\underset{\displaystyle O^-}{|}}{P^+}}-O-CH_2-\overset{\overset{\displaystyle H}{|}}{\underset{\underset{\displaystyle OH}{|}}{C}}-C\overset{\diagup O}{\diagdown_{O^-}} + 3\,H^+$$

RCHO RCO_2^-

The synthesis of ATP has a ΔG^0 of approximately $+7,000$ cal/mole, and therefore it is an "uphill" reaction.

$$ADP^{3-} + HPO_4^{2-} + H^+ \longrightarrow ATP^{4-} + H_2O$$

Addition of the two equations gives us the net result of the coupled reaction:[°]

$$RCHO + ADP^{3-} + HPO_4^{2-} \xrightarrow{2e} RCO_2^- + ATP^{4-} + 2\,H^+$$

$$\Delta G^\circ = -7,000 + 7,000 \approx 0\ \text{cal/mole}$$

Therefore, effectively all the free energy available from the oxidation reaction is transferred to the synthesis of ATP.

The complete mechanism for the oxidation is complex, but it is possible to approximate the key steps in the oxidation to illustrate the transfer of energy in the reaction. The nonenzymatic oxidation of the aldehyde in a water solution yields the acid directly, plus heat:

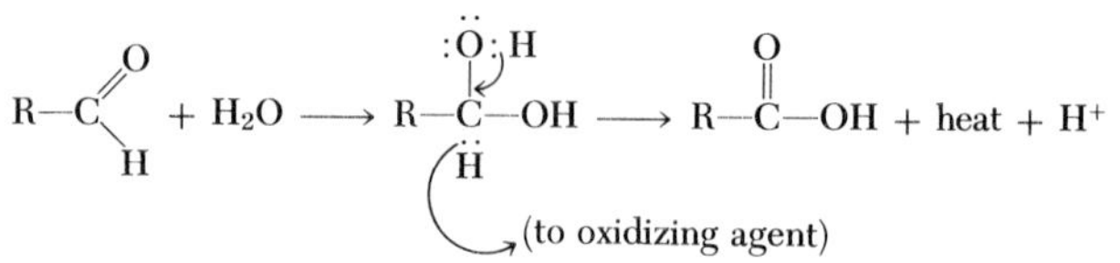

The reaction is straightforward, but it is wasteful of energy. In the enzymatic oxidation, an SH group on the enzyme first adds to the aldehyde; oxidation then gives a high-energy, sulfur-containing ester of the acid. Since it is a high-energy product, no energy is left to be wasted as heat:

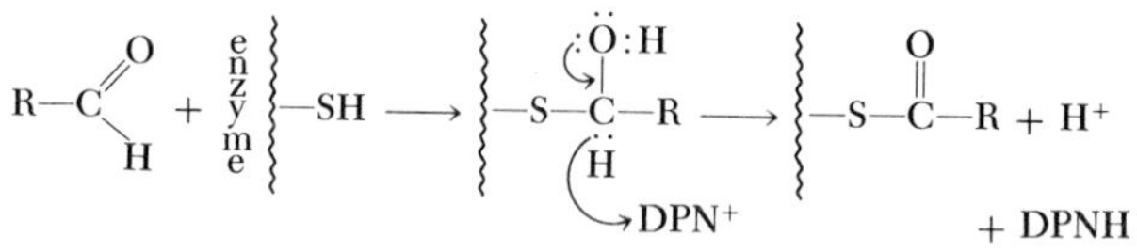

In subsequent steps, the high-energy ester reacts with phosphate ion to give an acyl phosphate, also a high-energy intermediate:

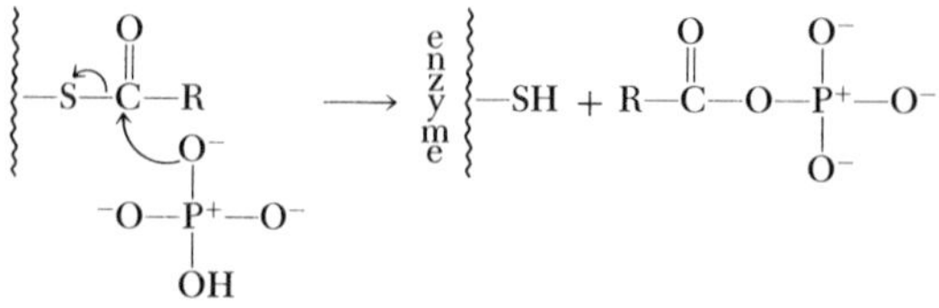

In the last step, ADP, also a phosphate, attacks the "active" acid to give ATP as well as the free carboxylic acid:

[°] The species shown are those that exist at pH 7. Thus,

$$RCO_2H \longrightarrow RCO_2^- + H^+ \quad \text{and} \quad H_3PO_4 \longrightarrow HPO_4^{2-} + 2H^+.$$

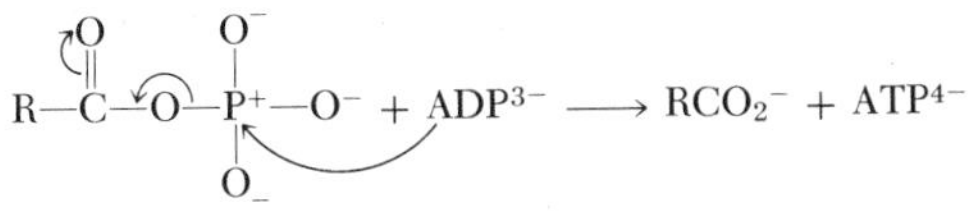

$$R-\overset{O}{\underset{}{C}}-O-\overset{O^-}{\underset{O_-}{P^+}}-O^- + ADP^{3-} \longrightarrow RCO_2^- + ATP^{4-}$$

It will be instructive to compare this specific instance of energy transfer along a chain of reactions with the graphic picture given in Figure 2.4 of the energy changes involved.

REACTIONS OF CARBOXYLIC ACIDS   Most of the reactions of carboxylic acids involve exchange of the OH group for some other group. The reactions usually involve the nucleophilic addition of a reagent to the carbonyl group, followed by an elimination reaction. For example, esters are readily formed in a mixture of a carboxylic acid and an alcohol:

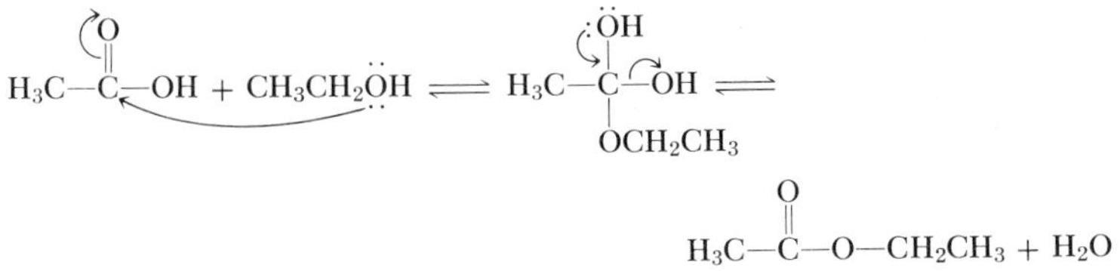

$$H_3C-\overset{O}{\underset{}{C}}-OH + CH_3CH_2\overset{..}{O}H \rightleftharpoons H_3C-\overset{\overset{..}{O}H}{\underset{OCH_2CH_3}{C}}-OH \rightleftharpoons$$

$$H_3C-\overset{O}{\underset{}{C}}-O-CH_2CH_3 + H_2O$$

Ethyl acetate

(Note that proton transfers are usually not indicated in reaction mechanisms; this is because they occur extremely rapidly relative to the bond-forming steps of the heavier atoms.) The concentration of ester at equilibrium is low if equal amounts of the reactants are used; high yields are obtained only if the water is removed during the course of the reaction, or if one of the reactants is present in excess. This reaction is also catalyzed by strong acids and a simple procedure for the preparation of esters involves the treatment of a mixture of the carboxylic acid and an excess of the alcohol with a small amount of concentrated sulfuric acid.

The addition-elimination mechanism outlined for this reaction has been confirmed through the use of the O^{18} isotope. When an alcohol labeled with O^{18} is condensed with a carboxylic acid, ordinary water and an ester containing all the O^{18} are obtained:

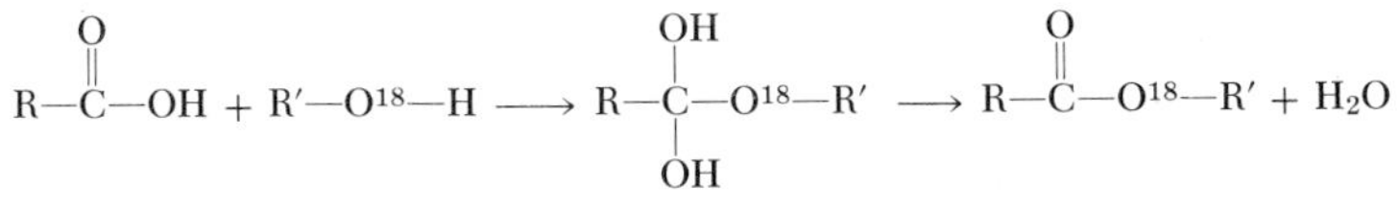

$$R-\overset{O}{\underset{}{C}}-OH + R'-O^{18}-H \longrightarrow R-\overset{OH}{\underset{OH}{C}}-O^{18}-R' \longrightarrow R-\overset{O}{\underset{}{C}}-O^{18}-R' + H_2O$$

The following hypothetical reaction course is, therefore, eliminated from consideration by the O^{18} results:

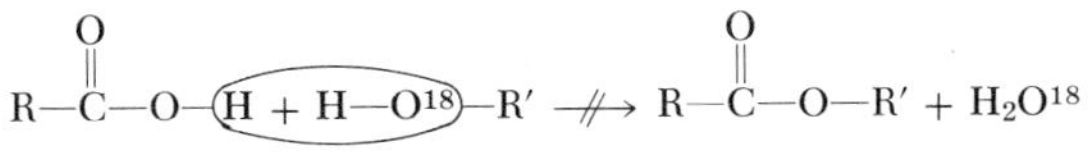

$$R-\overset{O}{\underset{}{C}}-O-(H + H-O^{18})-R' \not\longrightarrow R-\overset{O}{\underset{}{C}}-O-R' + H_2O^{18}$$

Esters are generally prepared, however, from acid anhydrides, the so-called "activated" forms of the carboxylic acid. Examples are given below:

$$H_3C-\overset{\overset{O}{\|}}{C}-O-\overset{\overset{O}{\|}}{C}-CH_3 \qquad H_3C-\overset{\overset{O}{\|}}{C}-O-\overset{\overset{O}{\|}}{C}-CH_2-CH_3$$

Acetic anhydride   *Acetic propionic anhydride*

$$H_3C-\overset{\overset{O}{\|}}{C}-Cl \qquad H_3C-\overset{\overset{O}{\|}}{C}-O-\overset{\overset{O}{\|}}{N}-O$$

Acetyl chloride   *Acetyl nitrate*

The carboxylic acid–carboxylic acid anhydrides are named by the addition of the word "anhydride" to the names of the acids involved. The anhydrides of inorganic acids are named essentially as salts with the organic radical (RCO) named by substituting the suffix *yl* for the *-ic* ending of the acids; that is, the *acetyl* radical is CH_3CO.

The anhydrides are readily prepared from the salt of one partner, and the acid chloride (RCOCl) of the other:

$$H_3C-\overset{\overset{O}{\|}}{C}-Cl + Na^+:\overset{..}{\underset{..}{O}}{}^- - \overset{\overset{O}{\|}}{C}-CH_2-CH_3 \longrightarrow \left[H_3C-\overset{\overset{:\overset{..}{O}:^- Na^+}{|}}{\underset{\underset{Cl}{|}}{C}}-O-\overset{\overset{O}{\|}}{C}-CH_2-CH_3 \right]$$

$$\longrightarrow H_3C-\overset{\overset{O}{\|}}{C}-O-\overset{\overset{O}{\|}}{C}-CH_2-CH_3 + Na^+Cl^-$$

The acid chlorides, in turn, are generally prepared by the reaction of the acids with thionyl chloride ($SOCl_2$):

$$H_3C-\overset{\overset{O}{\|}}{C}-\overset{..}{\underset{..}{O}}-H + :\overset{..}{Cl}-\overset{\overset{:\overset{..}{O}:}{|}}{S}-\overset{..}{Cl}: \longrightarrow HCl +$$

$$H_3C-\overset{\overset{O}{\|}}{C}-O-\overset{\overset{:\overset{..}{O}:}{|}}{\underset{\underset{:Cl:}{}}{S}} \longrightarrow H_3C-\overset{\overset{O}{\|}}{C}-Cl + SO_2$$

Acetyl chloride

The anhydrides are far more reactive than are the corresponding carboxylic acids or esters, for reasons that will be outlined at the end of Chapter 5. The anhydrides are generally used in the laboratory, therefore, when derivatives of the carboxylic acids are prepared:

$$CH_3\overset{\overset{O}{\|}}{C}-Cl + C_2H_5OH \longrightarrow CH_3\overset{\overset{O}{\|}}{C}-OC_2H_5 + HCl$$

$$H_3C-\overset{\overset{O}{\|}}{C}-O-\overset{\overset{O}{\|}}{C}-CH_3 + H_2\overset{..}{N}-OH \longrightarrow H_3C-\overset{\overset{O}{\|}}{C}-\overset{\overset{H}{|}}{N}-OH + CH_3CO_2H$$

Hydroxylamine   *Acetylhydroxylamine*

In the latter case, the corresponding acid itself (acetic acid) would give none of the desired product at room temperature. Instead, a simple acid-base reaction would occur to give a salt of the acid (an analog of ammonium acetate):

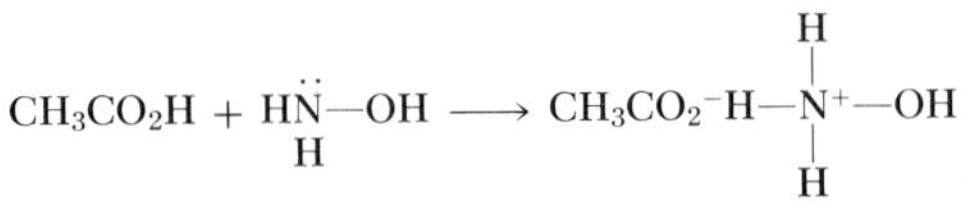

$$CH_3CO_2H + H\overset{..}{N}{-}OH \longrightarrow CH_3CO_2^- \; H{-}\overset{H}{\underset{H}{\overset{|}{N^+}}}{-}OH$$

Many of the reactions of carboxylic acids in biological systems involve mixed anhydrides with phosphoric acid, such as acetyl phosphate:

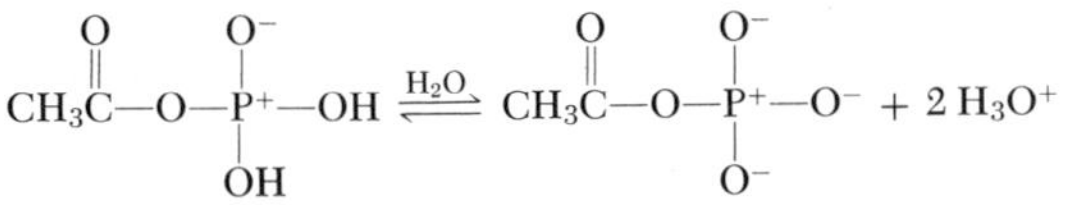

$$\underset{\text{Acetyl phosphate}}{CH_3\overset{O}{\overset{||}{C}}{-}O{-}\overset{O^-}{\underset{OH}{\overset{|}{\overset{|}{P^+}}}}{-}OH} \; \overset{H_2O}{\rightleftarrows} \; \underset{\substack{\textit{Common form} \\ \textit{at pH 7}}}{CH_3\overset{O}{\overset{||}{C}}{-}O{-}\overset{O^-}{\underset{O^-}{\overset{|}{\overset{|}{P^+}}}}{-}O^-} + 2\,H_3O^+$$

In some organisms acetyl phosphate is a precursor of acetyl coenzyme A (acetyl CoA; see p. 157), which is symbolized here as $CH_3\overset{}{\underset{O}{\overset{|}{C}}}{-}S{-}R$:

$$CH_3\overset{O}{\overset{||}{C}}{\frown}O{-}PO_3^{2-} + H{-}\overset{..}{\underset{..}{S}}{-}R \longrightarrow CH_3\overset{O}{\overset{||}{C}}{-}S{-}R + HPO_4^{2-}$$

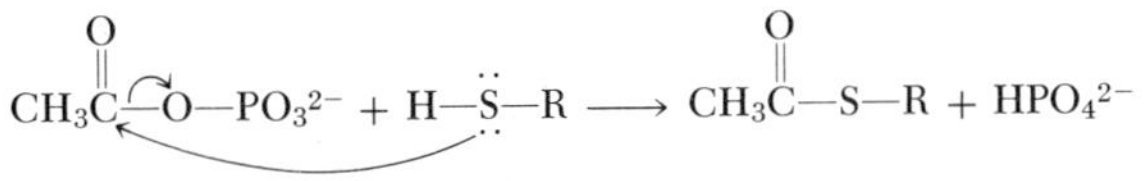

The acetyl group of acetyl CoA is now available for various condensations to give citric acid (p. 113, and see below), fatty acids, steroids, carotenes, and a number of other biologically important compounds.

The carbonyl group of acid derivatives is also subject to nucleophilic attack by carbon anions. The *Claisen condensation* is an example:

Step 1 $CH_3CO_2CH_3 + Na \longrightarrow Na^+ \; {}^-CH_2CO_2CH_3 + \tfrac{1}{2}\,H_2$

Step 2 $CH_3\overset{O}{\overset{||}{C}}{-}OCH_3 + {}^-CH_2\overset{O}{\overset{||}{C}}{-}OCH_3 \longrightarrow$

$$CH_3\overset{O^-}{\underset{\underset{OCH_3}{|}}{\overset{|}{C}}}{-}CH_2\overset{O}{\overset{||}{C}}{-}OCH_3 \longrightarrow CH_3\overset{O}{\overset{||}{C}}CH_2\overset{O}{\overset{||}{C}}{-}OCH_3 + CH_3O^-$$

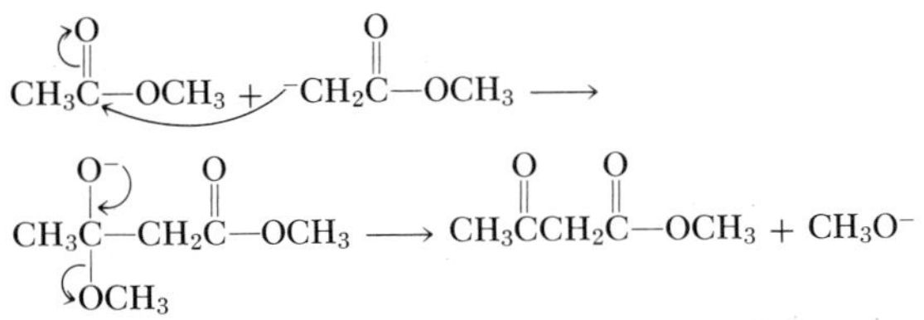

In this reaction, a new carbon–carbon bond is formed, thus lengthening the chain. Note the resemblance of this reaction to the aldol condensation discussed under aldehydes. A biological example of the Claisen condensation is the condensation of acetyl CoA with oxalacetic acid to give citric acid. This compound is a component of the citric acid cycle; it is an important source of energy and also of intermediates in biological reactions (shown next page).

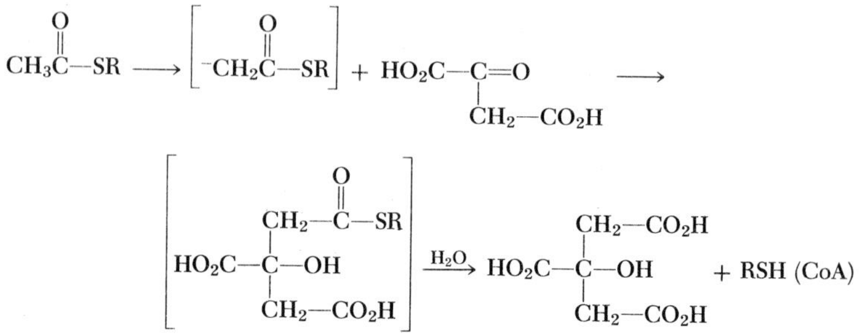

$$CH_3\overset{O}{\underset{\|}{C}}-SR \longrightarrow \left[^-CH_2\overset{O}{\underset{\|}{C}}-SR\right] + HO_2C-\underset{\underset{CH_2-CO_2H}{|}}{C}=O \longrightarrow$$

$$\left[\begin{array}{c} CH_2-\overset{O}{\overset{\|}{C}}-SR \\ HO_2C-C-OH \\ CH_2-CO_2H \end{array}\right] \xrightarrow{H_2O} \begin{array}{c} CH_2-CO_2H \\ HO_2C-C-OH \\ CH_2-CO_2H \end{array} + RSH\ (CoA)$$

Citric acid

Reduction of carboxylic acids The carboxyl group is reduced to the corresponding primary alcohol group by lithium aluminum hydride, as shown in the following reaction:

$$RCO_2H \xrightarrow{LiAlH_4} RCH_2OH$$

This is the only commonly used reagent that is able to reduce the carboxyl group. This fact can be turned to advantage when selective reductions are desired; the reactions in Figure 4.11 illustrate this point.

Decarboxylation of carboxylic acids Saturated, unsubstituted carboxylic acids are very stable compounds and they are decomposed only at high temperatures in the presence of a base:

$$H_3C-\overset{O}{\overset{\|}{C}}-O^-Na^+ \xrightarrow[300°]{NaOH} CH_4 + Na_2CO_3$$

Reactions of this type in which carboxylic acids yield carbon dioxide (or its derivatives) are called *decarboxylation* reactions.

Several kinds of organic acids undergo a much more facile decarboxylation than does acetic acid.

Figure 4.11 **The selective reduction of various functional groups in a complex carboxylic acid.**

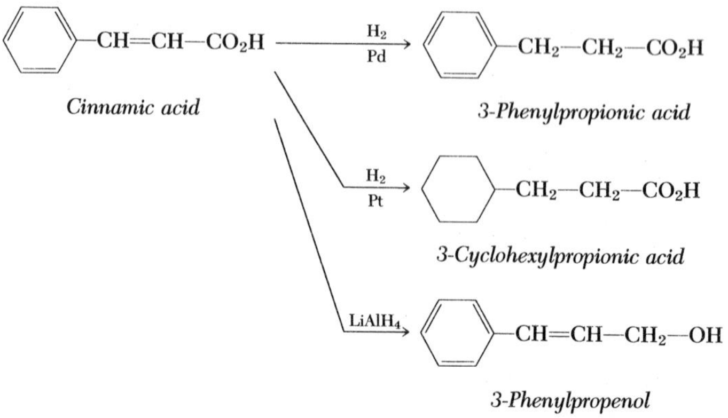

Cinnamic acid

3-Phenylpropionic acid

3-Cyclohexylpropionic acid

3-Phenylpropenol

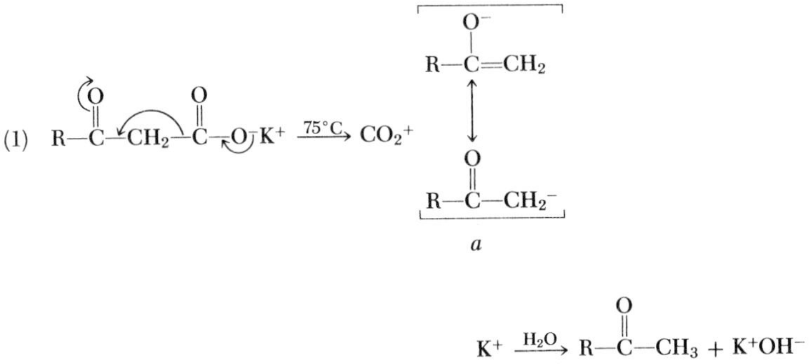

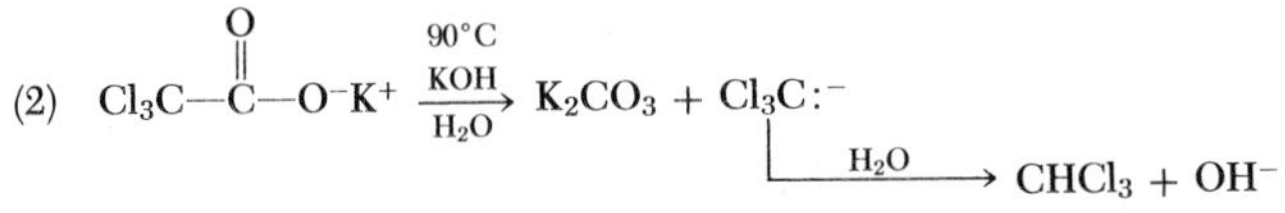

$$(1) \quad R-\overset{O}{\underset{\|}{C}}-CH_2-\overset{O}{\underset{\|}{C}}-O^-K^+ \xrightarrow{75°C} CO_2^+ \quad\quad \begin{bmatrix} \overset{O^-}{\underset{\|}{R-C=CH_2}} \\ \updownarrow \\ R-\overset{O}{\underset{\|}{C}}-CH_2^- \end{bmatrix}$$
$$a$$

$$K^+ \xrightarrow{H_2O} R-\overset{O}{\underset{\|}{C}}-CH_3 + K^+OH^-$$

$$(2) \quad Cl_3C-\overset{O}{\underset{\|}{C}}-O^-K^+ \xrightarrow[H_2O]{\overset{90°C}{KOH}} K_2CO_3 + Cl_3C:^- $$
$$\xrightarrow{H_2O} CHCl_3 + OH^-$$

Chloroform

In the first example, a carbon anion a is formed that is stabilized by resonance (electrons delocalized over the carbonyl group and the α carbon atom). In the second case, a carbon anion is formed that is stabilized by the electron-withdrawing inductive effect of the three chlorines. In contrast, the decarboxylation of acetate ion yields the high-energy, nonstabilized ion CH_3^-. It is the stabilization of the charge on the carbon anion intermediate that allows decarboxylations of types 1 and 2 to proceed readily. β-keto acids [for example, the reactant in equation (1)] are well known for their ability to decarboxylate. However, the double bond on carbon atom 3 is the important structural feature since it is involved in the resonance of the intermediate a. Thus, it is interesting that many types of biological decarboxylations depend on the introduction of a double bond in position 3 of the acid to facilitate the decarboxylation. Pyridoxal phosphate (symbolized R'CHO) is a coenzyme in one class of enzyme-catalyzed decarboxylations, the function of which is to introduce that double bond.

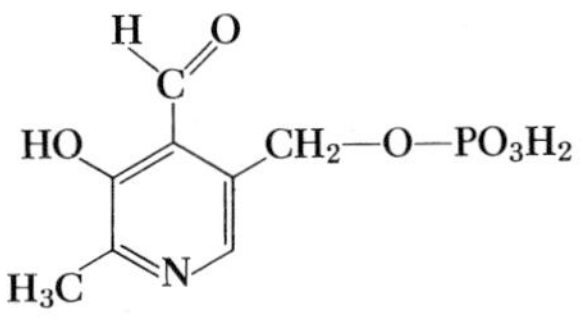

Pyridoxal phosphate

For example, the reaction sequence which is shown at the top of page 124 represents a plausible mechanism for the enzymatic decarboxylation of the amino acid alanine.

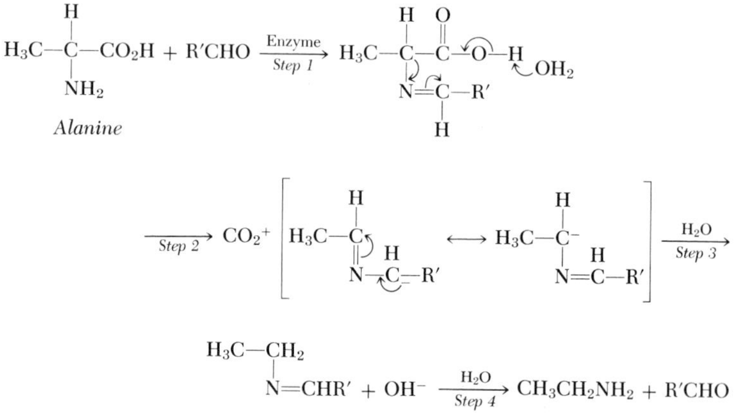

$$\underset{\substack{\text{Alanine}}}{\underset{\displaystyle \overset{\displaystyle H}{\underset{\displaystyle NH_2}{H_3C-C-CO_2H}}}{}} + R'CHO \xrightarrow[\text{Step 1}]{\text{Enzyme}} H_3C-\overset{\displaystyle H}{\underset{\displaystyle N=C-R'}{\underset{\displaystyle H}{C}}}-\overset{\displaystyle O}{C}\overset{\curvearrowleft}{-O-H}\curvearrowright OH_2$$

$$\xrightarrow[\text{Step 2}]{} CO_2^+ \left[H_3C-\overset{\displaystyle H}{\underset{\displaystyle N\curvearrowright C-R'}{\underset{\displaystyle H}{C}}} \longleftrightarrow H_3C-\overset{\displaystyle H}{\underset{\displaystyle N=C-R'}{\underset{\displaystyle H}{C^-}}} \right] \xrightarrow[\text{Step 3}]{H_2O}$$

$$\underset{\displaystyle N=CHR'}{\overset{\displaystyle H_3C-CH_2}{|}} + OH^- \xrightarrow[\text{Step 4}]{H_2O} CH_3CH_2NH_2 + R'CHO$$

In step 1, a typical carbonyl condensation occurs as outlined in Figure 4.8. In step 2, decarboxylation occurs to give the resonance-stabilized anion. Note that additional resonance contributors can be drawn showing that the negative charge is also delocalized over the pyridine ring; these additional forms would further stabilize the anion. The protonation in step 3 is a straightforward acid-base reaction, and the mechanism of step 4 is essentially a reverse of that of step 1. Pyridoxal phosphate (R'CHO) is regenerated in this last step to continue its catalytic work.

Esters ($R-\overset{\displaystyle O}{\overset{\|}{C}}-O-R'$) may be considered as alkyl derivatives of acids or alternatively as acid derivatives of alcohols. However, they are usually named as if the former were the case; that is, the name of the radical R' is given first, then a term is added for the acid portion, which is derived by dropping the acid ending, *-ic acid,* and substituting the suffix *-ate.* Selected examples of esters and their names are given in Figure 4.12.

It is interesting to note that whereas the lower-molecular-weight carboxylic acids have odors that range from objectionable (propionic) to vile (butanoic and pentanoic), the corresponding esters have very pleasant odors that resemble the odors of ripe fruits. For example, methyl butanoate has a pineapplelike odor and isoamyl acetate smells like ripe bananas. Mixtures of the lower-molecular-weight esters are in fact responsible for the flavors and aromas of many flowers and fruits. A number of interesting esters have been isolated recently from insect sources. The compound illustrated below is secreted by the female gypsy moth (*Porthetria dispar*). It is an incredibly potent attractant, and with the aid of this "insect perfume," a single female can attract males from a distance of up to 3 miles!

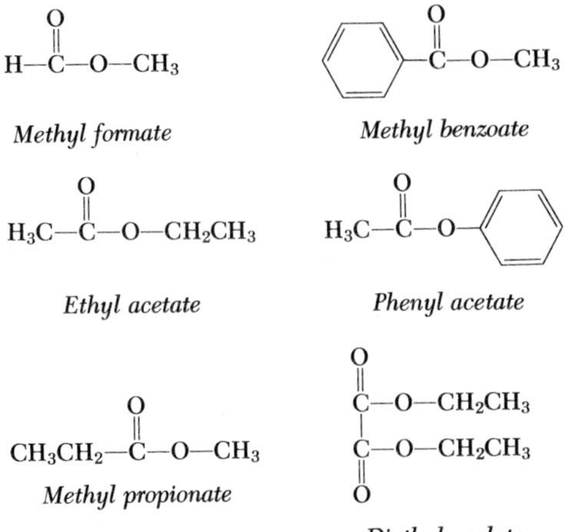

$$\underset{\textit{Methyl formate}}{H-\overset{\displaystyle O}{\overset{\|}{C}}-O-CH_3}$$

$$\underset{\textit{Methyl benzoate}}{\langle \bigcirc \rangle -\overset{\displaystyle O}{\overset{\|}{C}}-O-CH_3}$$

$$\underset{\textit{Ethyl acetate}}{H_3C-\overset{\displaystyle O}{\overset{\|}{C}}-O-CH_2CH_3}$$

$$\underset{\textit{Phenyl acetate}}{H_3C-\overset{\displaystyle O}{\overset{\|}{C}}-O-\langle \bigcirc \rangle}$$

$$\underset{\textit{Methyl propionate}}{CH_3CH_2-\overset{\displaystyle O}{\overset{\|}{C}}-O-CH_3}$$

$$\underset{\textit{Diethyl oxalate}}{\overset{\displaystyle O}{\overset{\|}{\underset{\underset{O}{\|}}{C}}}\overset{-O-CH_2CH_3}{\underset{-O-CH_2CH_3}{}}}$$

Figure 4.12   **Esters of carboxylic acids.**

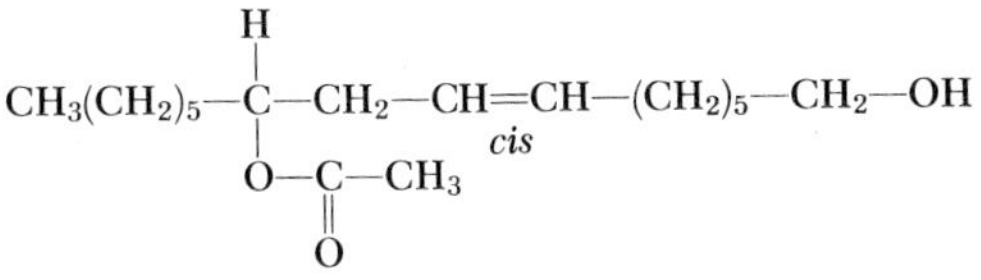

$$CH_3(CH_2)_5-\underset{\underset{\underset{O}{\|}}{\overset{\displaystyle |}{O-C-CH_3}}}{\overset{\displaystyle H}{\overset{\displaystyle |}{C}}}-CH_2-\underset{cis}{CH=CH}-(CH_2)_5-CH_2-OH$$

The gypsy moth sex attractant

PREPARATION OF THE ESTERS   Esters may be prepared directly from carboxylic acids as was outlined in the section on acids. They are usually prepared from the acid chlorides, however:

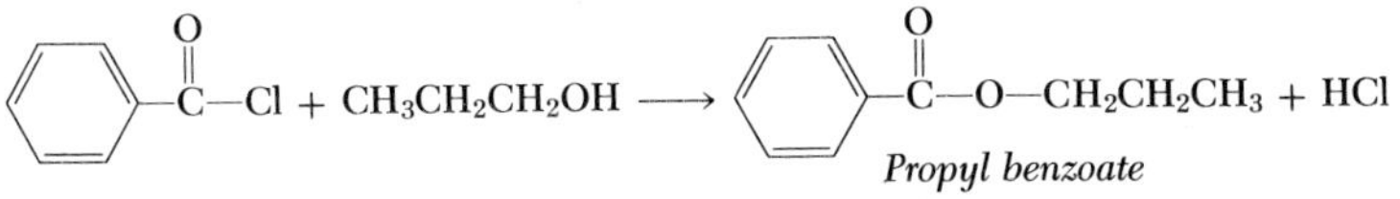

$$\langle \bigcirc \rangle -\overset{\displaystyle O}{\overset{\|}{C}}-Cl + CH_3CH_2CH_2OH \longrightarrow \underset{\textit{Propyl benzoate}}{\langle \bigcirc \rangle -\overset{\displaystyle O}{\overset{\|}{C}}-O-CH_2CH_2CH_3} + HCl$$

or from other esters by a process called *ester interchange:*

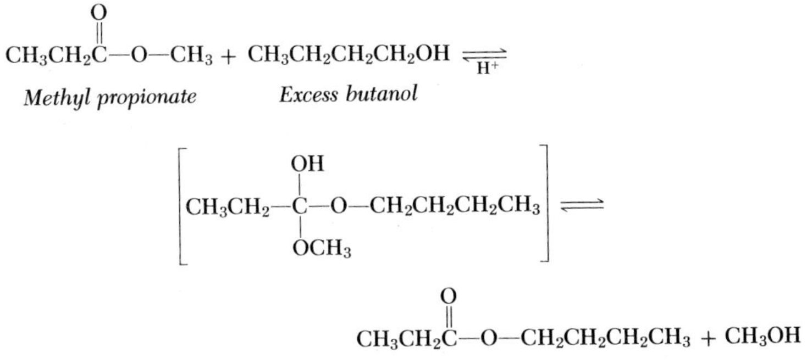

$$\underset{\textit{Methyl propionate}}{CH_3CH_2\overset{\displaystyle O}{\overset{\|}{C}}-O-CH_3} + \underset{\textit{Excess butanol}}{CH_3CH_2CH_2CH_2OH} \underset{H^+}{\rightleftharpoons}$$

$$\left[CH_3CH_2-\underset{\overset{\displaystyle |}{OCH_3}}{\overset{\overset{\displaystyle OH}{|}}{C}}-O-CH_2CH_2CH_2CH_3 \right] \rightleftharpoons$$

$$\underset{\textit{Butyl propionate}}{CH_3CH_2\overset{\displaystyle O}{\overset{\|}{C}}-O-CH_2CH_2CH_2CH_3} + CH_3OH$$

In this process, an equilibrium is established between two alcohols and the corresponding esters. If an excess of one alcohol is used or if the other alcohol is removed by distillation or some other technique, a good yield of the interchange ester (here the butyl propionate) can be obtained.

REACTIONS OF THE ESTERS   Most of the reactions of esters involve nucleophilic addition to the carbonyl group. Most esters react with water, for example, to yield the acid and alcohol from which the ester was made:

$$H_3C-\overset{\overset{O}{\|}}{C}-O-CH_3 + H_2O \rightleftharpoons \left[H_3C-\underset{\underset{OH}{|}}{\overset{\overset{OH}{|}}{C}}-O-CH_3 \right] \rightleftharpoons$$

$$H_3C-\overset{\overset{O}{\|}}{C}-OH + CH_3OH$$

The reaction is catalyzed by acids and the use of a large excess of water insures that the equilibrium will be shifted to the right; that is, that hydrolysis will be complete.

The steric effect   Esters with several alkyl groups on the α-carbon atom of the acid portion of the molecule react only very slowly with water. It is thought that this decrease in rate is due to a blocking of the approach of water molecules to the carbonyl group by the bulky alkyl groups (Figure 4.13). This interaction is called *steric hindrance,* or more generally, a *steric effect.* The base-catalyzed hydrolysis of esters of bulky alcohols, such as the acetate of tertiary butyl alcohol, is also abnormally slow, and the low rate is also attributed to the steric interaction of the nucleophile and the alkyl groups, but this time on the alcohol portion. We have not stressed the role of steric effects in organic reactions up to this point; however, it should be pointed out that when bulky groups near the reaction center can hinder the approach of a reagent, the resulting steric effects may outweigh both the inductive and resonance effects in determining the rate or course of a reaction.

Saponification of esters   Most esters react with aqueous bases to give alcohols and salts of carboxylic acids; this process is called *saponification:*

$$H_3C-\overset{\overset{O}{\|}}{C}-O-CH_2CH_3 + Na^+OH^- \rightleftharpoons \left[H_3C-\underset{\underset{OH}{|}}{\overset{\overset{O^-Na^+}{|}}{C}}-O-CH_2CH_3 \right] \longrightarrow$$

$$CH_3CO_2H + Na^+\,{}^-OCH_2CH_3 \longrightarrow CH_3CO_2^-Na^+ + CH_3CH_2OH$$

The reaction involves, in sequence, a nucleophilic attack of hydroxide ion on the carbonyl group to give the adduct, the regeneration of the carbonyl group to yield acetic acid and an alkoxide ion, and an acid-base reaction to yield acetate ion and the free alcohol. The saponification reaction is irreversible

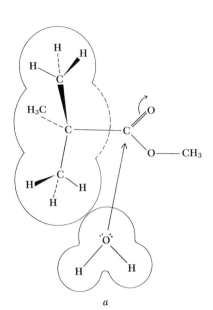

Figure 4.13 *Steric blocking of a nucleophile by the α-alkyl groups of a carboxylic ester. The sizes of the interacting species are approximate.*

since carboxylate ions do not react with nucleophiles (because of the repulsion of like charges):

$$CH_3-C\overset{O}{\underset{O^-}{\big\langle}} + OR^- \xrightarrow{\ \ //\ \ } H_3C-\overset{O^-}{\underset{OR}{\overset{|}{C}}}-O^-$$

Saponification is therefore a quantitative method for cleaving esters.

Esters also react with many of the nucleophiles that attack acid anhydrides, and often it is more convenient to prepare carboxylic acid derivatives from the esters than from the anhydrides.

$$H_3C-\overset{O}{\overset{\|}{C}}-O-CH_3 + :NH_3 \longrightarrow H_3C-\overset{O}{\overset{\|}{C}}-NH_2 + CH_3OH$$
Acetamide

$$H_3C-\overset{O}{\overset{\|}{C}}-O-CH_2CH_3 + \overset{..}{N}H_2-\overset{..}{N}H_2 \longrightarrow H_3C-\overset{O}{\overset{\|}{C}}-\overset{H}{\overset{|}{N}}-NH_2 + CH_3CH_2OH$$
Hydrazine   Acethydrazide

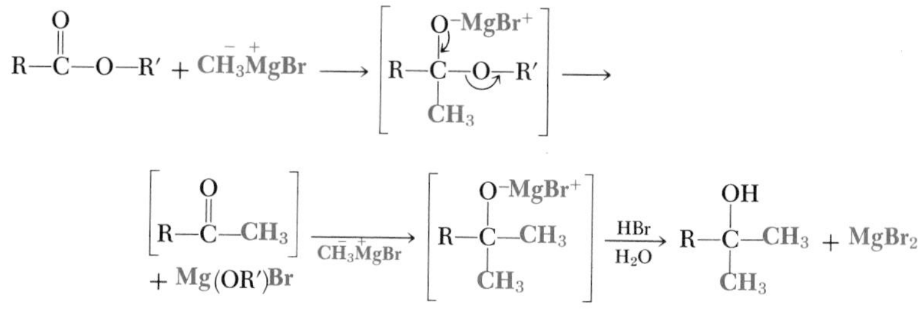

$$\underset{\substack{\text{O} \\ \|}}{\text{R—C—O—R'}} + \bar{\text{C}}\text{H}_3\overset{+}{\text{M}}\text{gBr} \longrightarrow \left[\underset{\substack{\text{CH}_3}}{\overset{\overset{\text{O}^-\text{MgBr}^+}{|}}{\text{R—C—O}\text{—R'}}} \right] \longrightarrow$$

$$\left[\underset{\substack{\text{O} \\ \|}}{\text{R—C—CH}_3} \right] \xrightarrow{\bar{\text{C}}\text{H}_3\overset{+}{\text{M}}\text{gBr}} \underset{\substack{\text{CH}_3}}{\overset{\overset{\text{O}^-\text{MgBr}^+}{|}}{\text{R—C—CH}_3}} \xrightarrow[\text{H}_2\text{O}]{\text{HBr}} \underset{\substack{\text{CH}_3}}{\overset{\overset{\text{OH}}{|}}{\text{R—C—CH}_3}} + \text{MgBr}_2$$
$$+ \text{Mg(OR')Br}$$

Figure 4.14 **The reaction of esters with a Grignard reagent.**

The reaction of Grignard reagents with esters This reaction is one of the better ways to synthesize alcohols (Figure 4.14). A magnesium derivative of the alcohol is the immediate product of the reaction; however, these compounds are readily hydrolyzed to the alcohols when water is added at the end of the reaction.

The reaction, in effect, converts an ester into a symmetrical alcohol. If R is an alkyl group, the product is a tertiary alcohol in which two of the alkyl groups came from the Gringnard reagent. The ketone is presumed to be an intermediate in this reaction, but since it reacts with the Grignard reagent at a faster rate than does the starting ester, it is rarely isolated from such reactions. Note that if the ester used is a derivative of formic acid (R=H), the product is a secondary alcohol (after treatment with water):

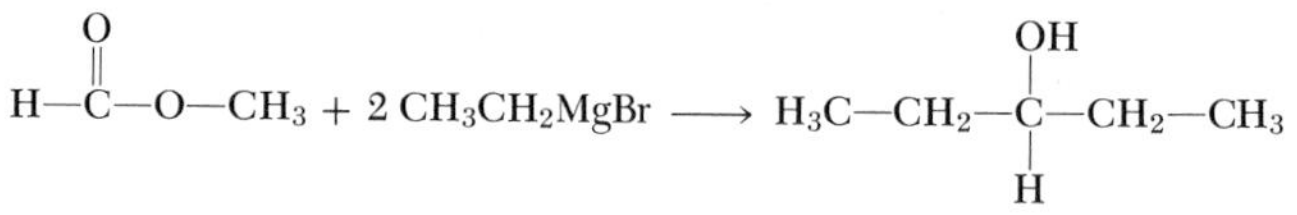

$$\underset{\substack{\text{O} \\ \|}}{\text{H—C—O—CH}_3} + 2\,\text{CH}_3\text{CH}_2\text{MgBr} \longrightarrow \underset{\substack{| \\ \text{H}}}{\overset{\overset{\text{OH}}{|}}{\text{H}_3\text{C—CH}_2\text{—C—CH}_2\text{—CH}_3}}$$

5 ORGANIC COMPOUNDS CONTAINING NITROGEN, SULFUR, AND PHOSPHORUS

COMPOUNDS OF NITROGEN

THE ORGANIC COMPOUNDS OF OXYGEN ARE CHARACTERIZED LARGELY BY THE acidic properties of the hydroxyl group, on the one hand, and by nucleophilic addition to the carbonyl group, on the other. Although there are analogs of these reactions in the chemistry of the nitrogen-containing compounds, their chemistry is usually dominated by the basicity of the nitrogen atom. The chemistry is more varied also because of the greater range of oxidation states available to nitrogen; the values range from -3 for ammonia and the amines to $+3$ for the nitro derivatives. Since the amines play a central role in the chemistry of the nitrogen compounds, our first section is devoted to these.

AMINES

Amines are derivatives of ammonia in which one or more of the hydrogen atoms are replaced by alkyl or aryl groups. If only one hydrogen is replaced (RNH_2), the compound is called a *primary* amine; if two hydrogens are replaced (R_2NH), the compound is called a *secondary* amine; and if three hydrogens are replaced (R_3N), the compound is called a *tertiary* amine. The amines are named as alkyl derivatives of ammonia in which the root term "*ammonia*"

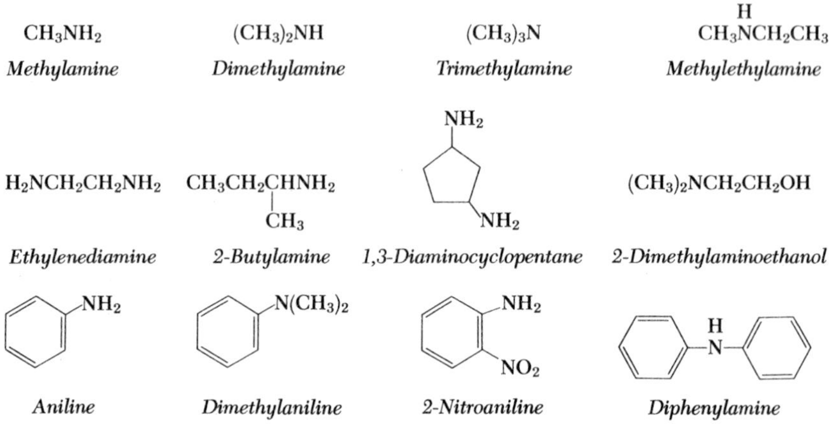
CH3NH2
Methylamine
(CH3)2NH
Dimethylamine
(CH3)3N
Trimethylamine
H
CH3NCH2CH3
Methylethylamine
H2NCH2CH2NH2
Ethylenediamine
CH3CH2CHNH2
CH3
2-Butylamine
NH2
NH2
1,3-Diaminocyclopentane
(CH3)2NCH2CH2OH
2-Dimethylaminoethanol
NH2
Aniline
N(CH3)2
Dimethylaniline
NH2
NO2
2-Nitroaniline
H
N
Diphenylamine

Figure 5.1 **Aliphatic and aromatic amines.**

is contracted to "amine," or the NH_2 group is treated as a substituent, in which case the prefixes *amino-*, *methylamino-*, and so on, are used. Many of the aromatic amines are referred to by their common names, however. Examples of the amines are given in Figure 5.1.

Ammonium ions are tetrahedrally symmetrical (Figure 5.2), and the hybridization of the nitrogen atom is presumed to be sp^3; in these respects, the ammonium ion resembles methane. Ammonia and the amines have approximately the same shape as does the ammonium ion, with the exception that an electron pair takes the place of one of the NH bonds; the hybridization of the nitrogen atom in these compounds may also be sp^3 (but a somewhat different interpretation of the hybridization is given in Chapter 1).

PROPERTIES OF THE AMINES The amines are more basic and less acidic than are the corresponding alcohols (as follows from the lower electronegativity of nitrogen relative to oxygen) and they are often used in reactions when weak bases are required. The basic ionization constants of the amines range from 10^{-3} to about 10^{-12}:

$$R_3N{:} + H{-}O{-}H \rightleftharpoons R_3\overset{+}{N}H\ OH^- \qquad K = \frac{[R_3\overset{+}{N}H][OH^-]}{[R_3N]}$$

The base strength of an amine is determined by the electron density on nitrogen,

Figure 5.2 **The geometry of the ammonium ion, ammonia, and trimethyl amine.**

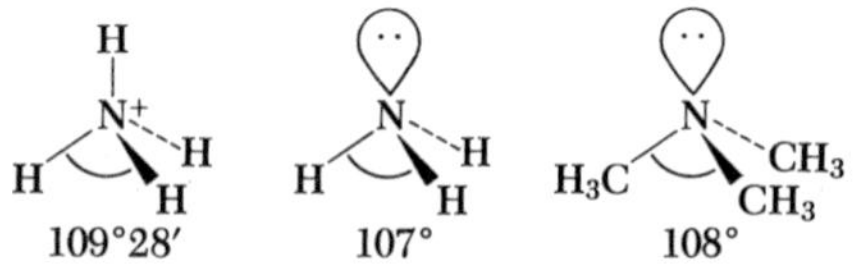
H
N+
H
H
H
109°28′
N
H
H
H
107°
N
H3C
CH3
CH3
108°

which in turn is determined by the nature of the substituent R. Since alkyl groups are electron-releasing groups, most alkylamines are stronger bases than is ammonia itself. Electron-withdrawing groups decrease the base strength; for example, $CF_3CH_2NH_2$ is a relatively weak base because the inductive effect of the fluorine atoms decreases the electron density on nitrogen. Aromatic amines such as aniline are weaker bases than ammonia because of resonance interactions; the unshared electron pair is not localized on the nitrogen atom as it is in ammonia, but it is delocalized over the aromatic ring.

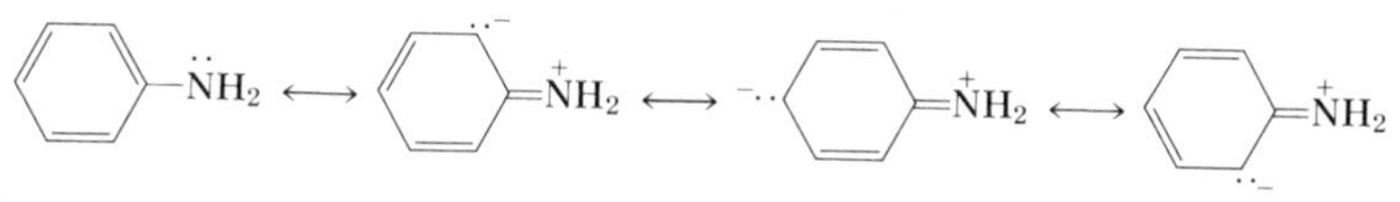

This resonance interaction is lacking in the ammonium salt,

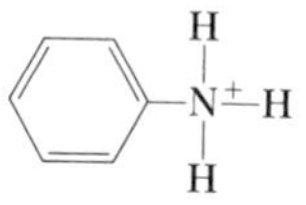

and consequently considerable resonance energy is lost when the amine is converted into the ion. For this reason, and because of the low electron density on nitrogen, aromatic amines tend to remain in the unprotonated form; that is to say, they are weak bases.

The resonance interaction of aromatic amines is reflected in the high reactivity of these compounds to electrophilic substitution and in the orientation observed. Aniline, for example, reacts extremely rapidly with bromine to give 2,4,6-tribromoaniline:

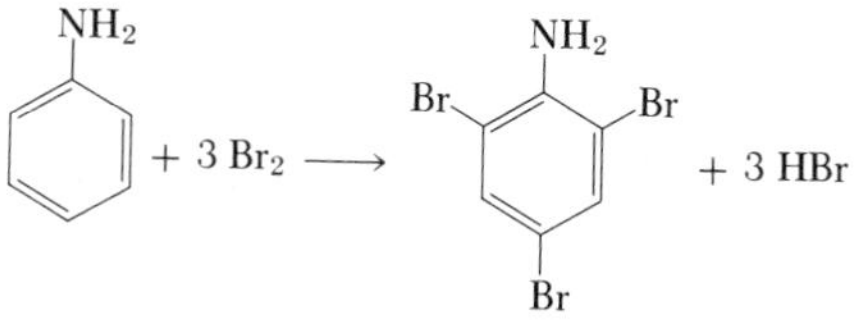

whereas toluene ($C_6H_5CH_3$) under the same reaction conditions gives no reaction whatsoever (p-bromotoluene is formed, however, when the reaction is catalyzed by $FeBr_3$ and is carried out at higher temperatures). The halogenated anilines are very insoluble in water and this reaction with bromine serves as a characteristic test for solutions of the aromatic amines. In strongly acidic solutions, aniline exists as the anilinium salt; under these conditions aniline does not react with bromine and when nitrated under forcing conditions, it gives meta-nitroaniline rather than the ortho and para isomers (the positively charged NH_3^+ group that is formed in acidic solutions is an electron-attracting group).

A list of ionization constants for various amines (Table 5.1) illustrates the effect of substituents on the base strength. In the aromatic amine series,

Table 5.1   *Base strengths of amines*

AMINE	K	
Dimethylamine	6.0×10^{-4}	
Methylamine	5.1×10^{-4}	
Trimethylamine	6.3×10^{-5}	
Ammonia	1.8×10^{-5}	Base strength
4-Methylaniline	1.2×10^{-9}	
Aniline	4.2×10^{-10}	
4-Chloroaniline	1.5×10^{-10}	
4-Nitroaniline	1.0×10^{-13}	

electron-attracting groups clearly decrease the base strength and electron-donating groups increase the base strength. In the aliphatic series, the base strength is proportional to the number of alkyl groups attached to nitrogen but only up to a point. It will be observed that although methylamine is a stronger base than ammonia, and dimethylamine is a stronger base than methylamine, trimethylamine is weaker than dimethylamine. This weakening of base strength has been observed in other highly substituted amines and it has been attributed to a steric effect of the bulky alkyl groups attached to the nitrogen.

It should be pointed out that the formation of hydroxide ions by amines in water is not necessary for these compounds to show acid-base reactions: $CH_3NH_2 + H_2O \rightleftharpoons CH_3NH_3^+ OH^-$. The amines themselves are bases (Lewis bases—by the definition given in Chapter 2). Ethylamine, for example, reacts with hydrogen chloride in an organic solvent to give ethylammonium chloride by a direct proton transfer:

$$CH_3CH_2N: \quad H:Cl \longrightarrow CH_3CH_2NH_3^+ Cl^-$$

Full proton transfer does not occur when a weaker acid such as ethanol is used, however:

$$CH_3CH_2\ddot{N}H_2 + CH_3CH_2OH \longrightarrow CH_3CH_2N:\text{---}HOCH_2CH_3$$
$$X$$

As we have seen from our discussion of the properties of alcohols, species X represents an ethanol molecule hydrogen-bonded to ethylamine. Since the amine is a stronger base than ethanol, this type of hydrogen bond is stronger than that which occurs in pure ethanol:

$$CH_3CH_2\text{---}\ddot{O}:\text{---}H\ddot{O}CH_2CH_3.$$

Pure amines also form hydrogen bonds,

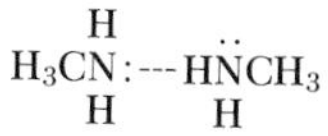

$$\begin{array}{ccc} & H & \ddot{} \\ H_3CN & : \text{---} & H\overset{..}{N}CH_3 \\ & H & H \end{array}$$

but since the amines are very weak acids, these hydrogen bonds are weak.

PREPARATION OF THE AMINES   Most aliphatic amines may be prepared by the displacement of the halide ions from alkyl halides by ammonia and the amines:

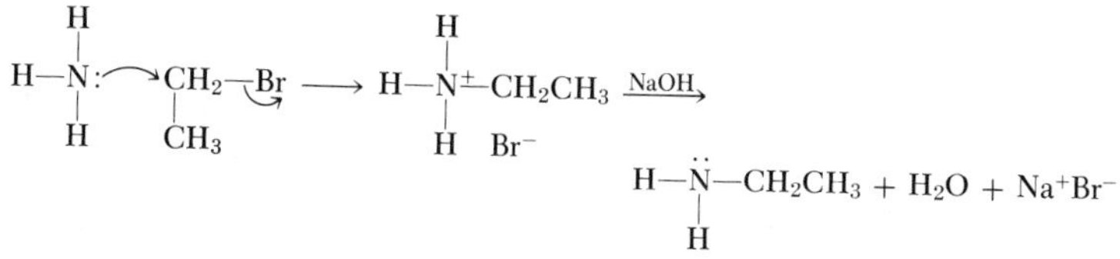

$$\begin{array}{c} H \\ | \\ H-N: \curvearrowright CH_2-Br \\ | \\ H \quad CH_3 \end{array} \longrightarrow \begin{array}{c} H \\ | \\ H-N^{\pm}-CH_2CH_3 \\ | \\ H \quad Br^- \end{array} \xrightarrow{\text{NaOH}}$$

$$\begin{array}{c} \ddot{} \\ H-N-CH_2CH_3 + H_2O + Na^+Br^- \\ | \\ H \end{array}$$

Ethylamine

A substituted ammonium salt is formed as an intermediate in the reaction but the free amine may be obtained from this salt by treatment with sodium hydroxide. A few examples are listed to illustrate the range and versatility of the reaction.

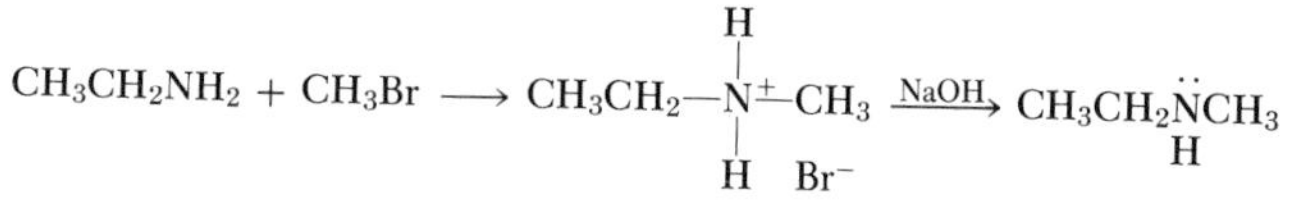

$$CH_3CH_2NH_2 + CH_3Br \longrightarrow \begin{array}{c} H \\ | \\ CH_3CH_2-N^{\pm}-CH_3 \\ | \\ H \quad Br^- \end{array} \xrightarrow{\text{NaOH}} \begin{array}{c} CH_3CH_2\ddot{N}CH_3 \\ | \\ H \end{array}$$

Methylethylamine

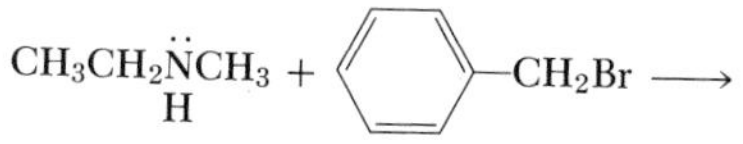

$$\begin{array}{c} CH_3CH_2\ddot{N}CH_3 \\ | \\ H \end{array} + \bigcirc\hspace{-1em}\text{—}CH_2Br \longrightarrow$$

α-Bromotoluene

$$\begin{array}{c} H \\ | \\ CH_3CH_2-N^{\pm}-CH_2 \\ | \\ CH_3 \quad Br^- \end{array}\bigcirc \xrightarrow{\text{NaOH}} \begin{array}{c} \ddot{} \\ CH_3CH_2-N-CH_2 \\ | \\ CH_3 \end{array}\bigcirc$$

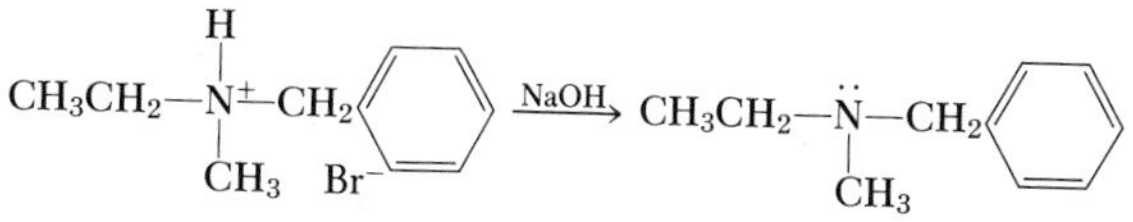

Methylethyl(α-tolyl)amine

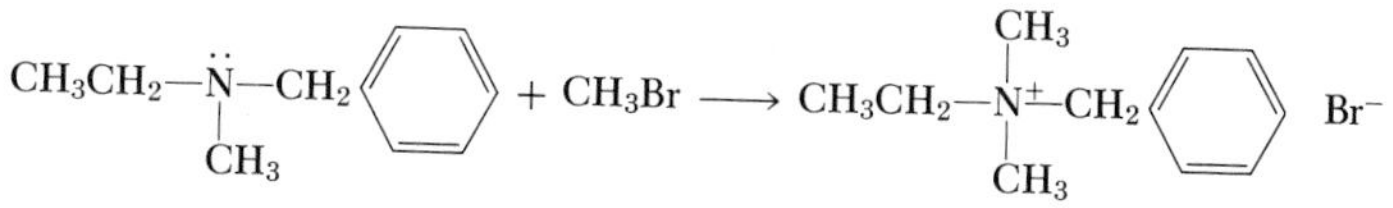

$$\begin{array}{c} \ddot{} \\ CH_3CH_2-N-CH_2 \\ | \\ CH_3 \end{array}\bigcirc + CH_3Br \longrightarrow \begin{array}{c} CH_3 \\ | \\ CH_3CH_2-N^{\pm}-CH_2 \\ | \\ CH_3 \end{array}\bigcirc Br^-$$

Dimethylethyl(α-tolyl)ammonium
bromide

In the last example, a tertiary amine has been converted into a *quaternary salt*. The quaternary salts are neutral, fully ionic compounds that have the properties of typical inorganic salts. Most primary and secondary amines may be transformed directly into quaternary salts if an excess of an alkyl halide and sodium hydroxide is used since the hydroxide converts the ammonium salts into the free amines, which are realkylated, and so on until the quaternary stage is reached.

$$CH_3CH_2NH_2 + CH_3Br \xrightarrow{NaOH} CH_3CH_2-\overset{\overset{\displaystyle CH_3}{|}}{\underset{\underset{\displaystyle CH_3}{|}}{N^{\pm}}}-CH_3 \quad Br^-$$
$$(excess)$$

Most of the simpler biological amines are formed by the decarboxylation of the amino acids, a process outlined on p. 122. Thus, serine yields ethanolamine (*a*) on decarboxylation. This compound in turn is converted into choline (*b*) by enzymatic methylation with the amino acid methionine as the source of methyl groups:

$$H_2NCH_2CH_2OH \qquad (CH_3)_3\overset{+}{N}CH_2CH_2OH \qquad (CH_3)_3\overset{+}{N}CH_2CH_2OCOCH_3$$
$$a \qquad\qquad\qquad b \qquad\qquad\qquad\qquad c$$

The methyl transfer is outlined on p. 153, but it seems certain that the transfer of methyl groups must be very similar to the nucleophilic displacement reactions used in the laboratory for alkylating amines (discussed above). The acetyl derivative of choline (*c*) is a part of the chemical chain involved in the transmission of nerve impulses at most nerve ends. It is synthesized in turn from an esterification of choline by the "activated form" of acetic acid, acetyl CoA (p. 157).

The Hofmann rearrangement   This reaction is one of the most reliable of the laboratory methods used for the preparation of primary amines. The following sequence is involved in the reaction, although technically the term "Hofmann rearrangement" refers only to the last step:

$$R-\overset{\overset{\displaystyle O}{\|}}{C}-OH \xrightarrow{SOCl_2} R-\overset{\overset{\displaystyle O}{\|}}{C}-Cl \xrightarrow{NH_3} R-\overset{\overset{\displaystyle O}{\|}}{C}-NH_2 \xrightarrow[\substack{NaOH \\ H_2O}]{Br_2} R-NH_2 + CO_2$$

The reaction, in essence, converts a carboxylic acid into an amine with the overall loss of one carbon atom. If the product contains a primary alkyl group, as in RCH_2NH_2, it can be oxidized with potassium permanganate to form a new carboxylic acid,

$$RCH_2NH_2 \xrightarrow{KMnO_4} RCO_2H$$

and the sequence can be repeated. For this reason, the Hofmann rearrangement and the oxidation step are often used to degrade long-chain carboxylic acids, carbon atom by carbon atom.

The mechanism of this reaction involves several intermediates (Figure 5.3). The first step is the nucleophilic displacement of bromide ion to give the

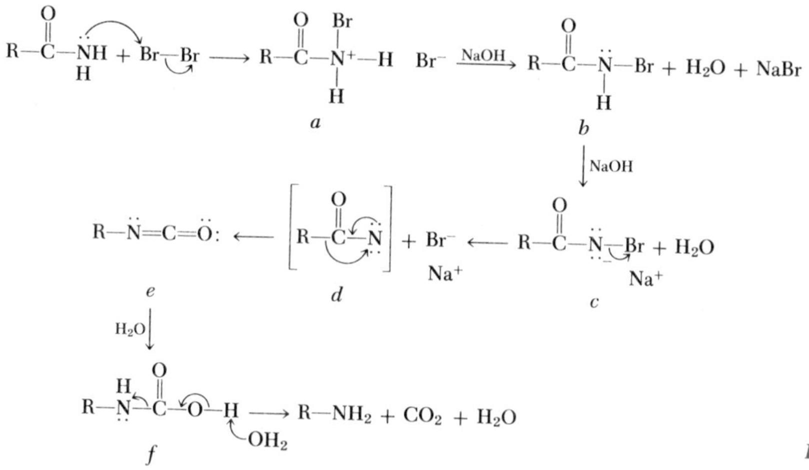

$$R-\overset{\overset{\displaystyle O}{\|}}{C}-\overset{\overset{}{}}{\underset{H}{\ddot{N}}}H + Br-Br \longrightarrow R-\overset{\overset{\displaystyle O}{\|}}{C}-\overset{\overset{Br}{|}}{\underset{H}{N^+}}-H \quad Br^- \xrightarrow{\text{NaOH}} R-\overset{\overset{\displaystyle O}{\|}}{C}-\overset{\overset{}{}}{\underset{H}{\ddot{N}}}-Br + H_2O + NaBr$$

$$a \qquad\qquad\qquad b$$

$$\Big\downarrow \text{NaOH}$$

$$R-\ddot{N}=C=\ddot{O}: \longleftarrow \left[R-\overset{\overset{\displaystyle O}{\|}}{C}-\ddot{N} \right] + Br^- \longleftarrow R-\overset{\overset{\displaystyle O}{\|}}{C}-\ddot{N}-Br + H_2O$$

$$e \qquad\qquad d \qquad Na^+ \qquad\qquad c \qquad Na^+$$

$$\Big\downarrow H_2O$$

$$R-\overset{\overset{\displaystyle H}{\diagdown}}{N}-\overset{\overset{\displaystyle O}{\|}}{C}-O-H \longrightarrow R-NH_2 + CO_2 + H_2O$$

$$f \qquad\qquad \overset{}{\underset{OH_2}{\diagup}}$$

Figure 5.3 **The mechanism of the Hofmann rearrangement.**

substituted ammonium salt (*a*); this compound then reacts with a base to give the free *N*-bromoamide (*b*). Because of the presence of two electron-withdrawing groups, this compound is an acid, and reaction with base gives the negative ion (*c*). The loss of a bromide ion then gives the highly reactive monovalent species (*d*), which electronically resembles a carbonium ion. The alkyl group and its pair of electrons move to the electron-deficient nitrogen to give an intermediate (*e*) called an *isocyanate*. The addition of water across the carbon-nitrogen double bond yields a compound (*f*) with a carboxyl group attached directly to nitrogen. These compounds are called *carbamic acids* and although their esters are stable, the carbamic acids themselves readily undergo decarboxylation to give the free amine. It can be seen that although the Hofmann rearrangement as a whole is complex, the individual steps are reasonably simple.

Isocyanates Isocyanates may be prepared by a number of reactions that permit the isolation of the pure compounds. A direct method involves the nucleophilic displacement of halide ions from the alkyl halides by cyanate ion:

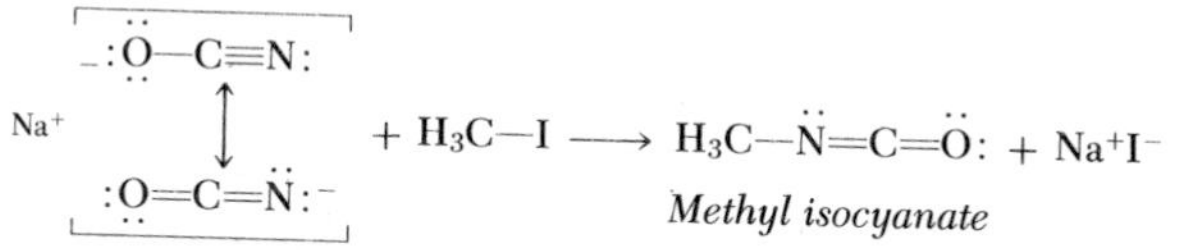

$$Na^+ \quad \begin{bmatrix} \_:\ddot{O}-C\equiv N: \\ \Big\updownarrow \\ :\ddot{O}=C=\ddot{N}:^- \end{bmatrix} + H_3C-I \longrightarrow H_3C-\ddot{N}=C=\ddot{O}: + Na^+I^-$$

Methyl isocyanate

Sodium cyanate

A second method, which is especially valuable for the preparation of aromatic isocyanates, involves the reaction of phosgene (the acid chloride of carbonic acid) with primary amines (shown at the top of page 136).

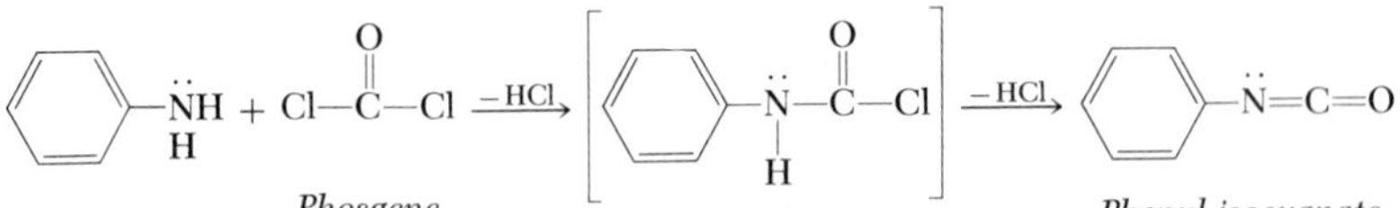

$$\text{C}_6\text{H}_5\text{—}\ddot{\text{N}}\text{H} + \text{Cl—}\overset{\displaystyle \text{O}}{\overset{\|}{\text{C}}}\text{—Cl} \xrightarrow{-\text{HCl}} \left[\text{C}_6\text{H}_5\text{—}\overset{..}{\underset{\text{H}}{\text{N}}}\text{—}\overset{\displaystyle \text{O}}{\overset{\|}{\text{C}}}\text{—Cl} \right] \xrightarrow{-\text{HCl}} \text{C}_6\text{H}_5\text{—}\ddot{\text{N}}\text{=C=O}$$

Phosgene   *Phenyl isocyanate*

The isocyanates react with alcohols to give esters of carbamic acid, in a similar fashion to their reaction with water (Figure 5.3):

$$\text{H}_3\text{C—}\ddot{\text{N}}\text{=C=}\ddot{\text{O}}: + \text{CH}_3\text{OH} \longrightarrow \text{H}_3\text{C—}\overset{\text{H}}{\underset{|}{\text{N}}}\text{—}\overset{\displaystyle \text{O}}{\overset{\|}{\text{C}}}\text{—O—CH}_3$$

Methyl-N-methylcarbamate

In addition, they react with ammonia or the amines to give derivatives of *urea* (a compound that may be considered the diamide of carbonic acid):

$$\text{CH}_3\ddot{\text{N}}\text{=C=}\ddot{\text{O}}: + :\text{NH}_3 \longrightarrow \text{H}_3\text{C—}\overset{\text{H}}{\underset{|}{\text{N}}}\text{—}\overset{\displaystyle \text{O}}{\overset{\|}{\text{C}}}\text{—}\overset{\text{H}}{\underset{|}{\text{N}}}\text{H}$$

N-Methylurea

The isocyanates are not isolable intermediates in biological processes, probably because of their ready hydrolysis. The more stable isothiocyanates (R—N=C=S), however, have been isolated from mustard, cabbage, and the other *Cruciferae*, where they are responsible for the pungent odor of these plants. Carbamic acid itself is an intermediate in the excretion of ammonia from the body, being ultimately converted into urea.

$$\text{H}_3\text{N}: + \text{O=C=O} \longrightarrow \text{H}_2\text{N—CO}_2\text{H} \rightarrow \rightarrow \rightarrow \text{H}_2\text{N—}\overset{\displaystyle \text{O}}{\overset{\|}{\text{C}}}\text{—NH}_2$$

Carbamic acid   *Urea*

Preparation of amines by reduction   The catalytic reduction of nitro compounds is a useful method for the preparation of primary amines. It is particularly valuable in the synthesis of aromatic amines since the nitro derivatives of the aromatic compounds are readily made by direct nitration, as shown below:

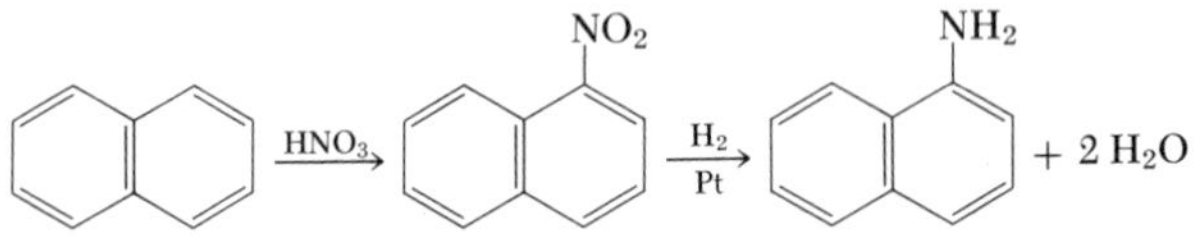

1-Nitronaphthalene

The reduction of nitriles and amides with lithium aluminum hydride, a reagent which, it will be recalled, is used to reduce acids to the corresponding alcohols, is another versatile method for the synthesis of amines:

$$H_3C-C\equiv N \xrightarrow{\text{LiAlH}_4} CH_3CH_2NH_2$$

$$H_3C-\overset{\overset{\displaystyle O}{\|}}{C}-NH_2 \xrightarrow{\text{LiAlH}_4} CH_3CH_2NH_2$$

$$H_3C-\overset{\overset{\displaystyle O}{\|}}{C}-N(CH_3)_2 \xrightarrow{\text{LiAlH}_4} CH_3CH_2N(CH_3)_2$$

REACTIONS OF THE AMINES   We have already mentioned the acid-base reactions of amines and also the reaction of amines with alkyl halides. The latter reaction is related to a useful method for the synthesis of olefins called the Hofmann elimination reaction.

THE HOFMANN ELIMINATION REACTION   Amines of all types react with an excess of methyl iodide in the presence of base to give a quaternary salt:

$$CH_3CH_2\overset{\overset{\displaystyle H}{|}}{N}CH_2CH_3 + 2\ CH_3I \xrightarrow{\text{KOH}} CH_3CH_2-\overset{\overset{\displaystyle CH_3}{|}}{\underset{\underset{\displaystyle CH_3}{|}}{N^+}}-CH_2CH_3\quad I^-$$

These compounds react with silver hydroxide to give a precipitate of silver iodide and a solution of a quaternary hydroxide:

$$CH_3CH_2NR_3{}^+I^- + AgOH \longrightarrow CH_3CH_2NR_3{}^+OH^- + AgI$$

The quaternary hydroxides are strong bases, equal in strength to the alkali metal hydroxides, and when they are heated, an elimination reaction involving the hydroxide ion occurs in which a tertiary amine, water, and an alkene are formed:

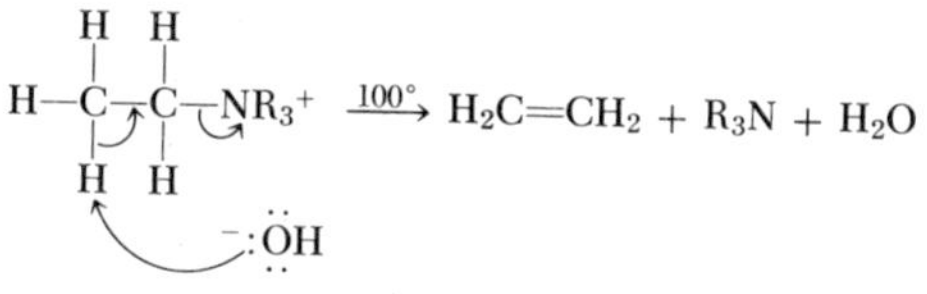

$$H-\overset{\overset{\displaystyle H}{|}}{\underset{\underset{\displaystyle H}{|}}{C}}-\overset{\overset{\displaystyle H}{|}}{\underset{\underset{\displaystyle H}{|}}{C}}-NR_3{}^+ \xrightarrow{100°} H_2C{=}CH_2 + R_3N + H_2O$$

$$-:\overset{..}{\underset{..}{O}}H$$

This reaction is a useful and general method for the synthesis of alkenes; the only structural requirement is that the amine must possess a β hydrogen atom.

The Hofmann elimination reaction is often used to prove the structures of complex amines, as shown in the following sequence:

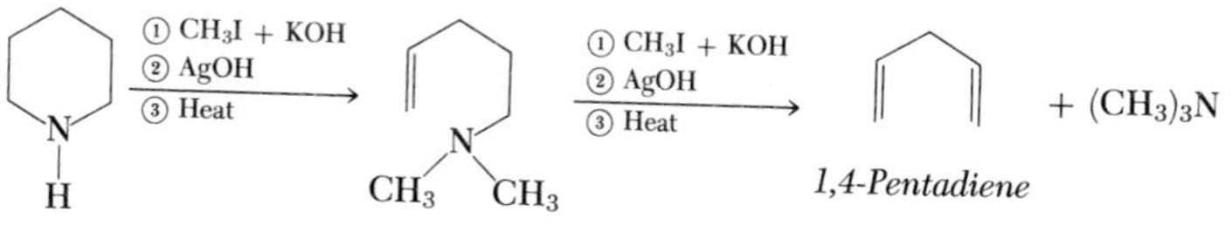

Piperidine   CH$_3$ CH$_3$ 1,4-Pentadiene   + (CH$_3$)$_3$N

The research chemist, at this point, might elect to oxidize the pentadiene to malonic acid ($HO_2CCH_2CO_2H$), a crystalline compound, which would then

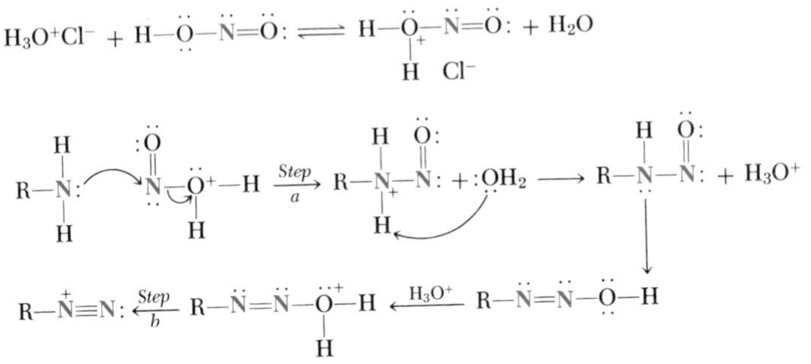

$$\text{H}_3\text{O}^+\text{Cl}^- + \text{H} - \overset{..}{\underset{..}{\text{O}}} - \overset{..}{\text{N}} = \overset{..}{\text{O}}: \rightleftharpoons \text{H} - \underset{\underset{\text{H}}{|}\quad\text{Cl}^-}{\overset{+}{\text{O}}} - \overset{..}{\text{N}} = \overset{..}{\text{O}}: + \text{H}_2\text{O}$$

Figure 5.4   *The reaction of nitrous acid with primary amines to give diazonium salts.*

be compared with an authentic sample of the acid. From his knowledge of these reactions, and from the identification of malonic acid, he would know that the starting material was piperidine.

DIAZONIUM IONS AND AZO DYES   Aromatic primary amines react with nitrous acid (generated from sodium nitrite [NaNO$_2$] and a strong acid such as HCl) to give an ionic species called a diazonium salt:

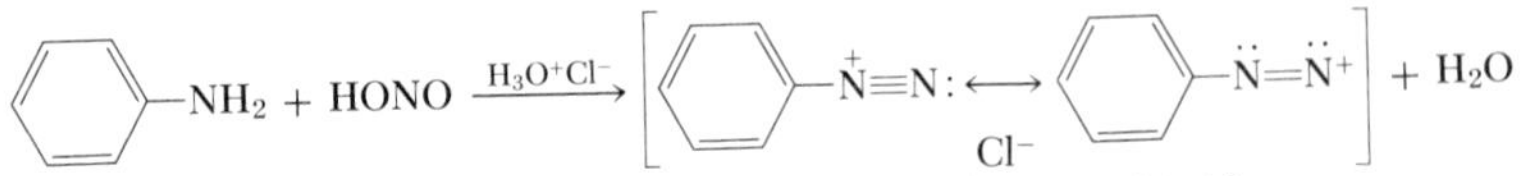

A number of steps are involved in this reaction (Figure 5.4). Step *a* is a displacement reaction and step *b* is an ionization reaction, whereas the remaining steps involve simple proton transfers.

The diazonium salts undergo a number of displacement reactions which are of considerable use in organic synthesis (Figure 5.5).

Figure 5.5   *Reactions of aromatic diazonium salts.*

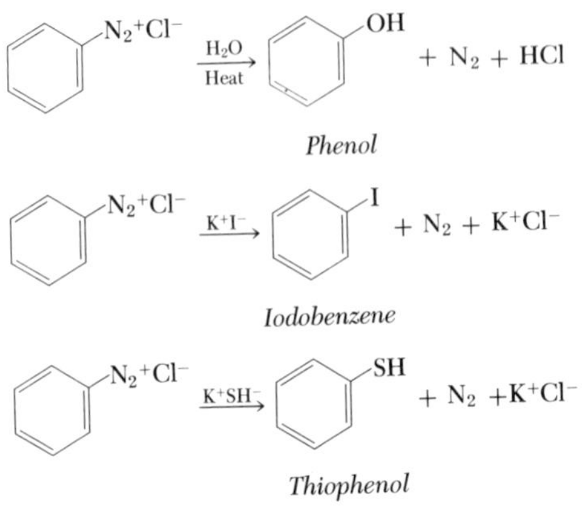

An interesting reaction in which nitrogen is not lost is the so-called coupling reaction, in which diazonium salts react with phenols and with aromatic amines by electrophilic substitution to give derivatives containing the azo linkage (—N=N—), which are known as *azo* compounds:

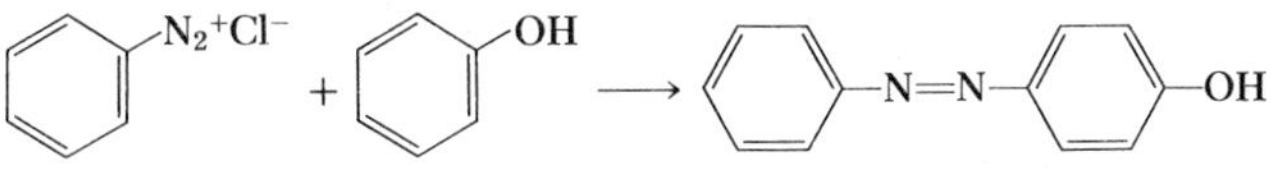

4-Hydroxyazobenzene

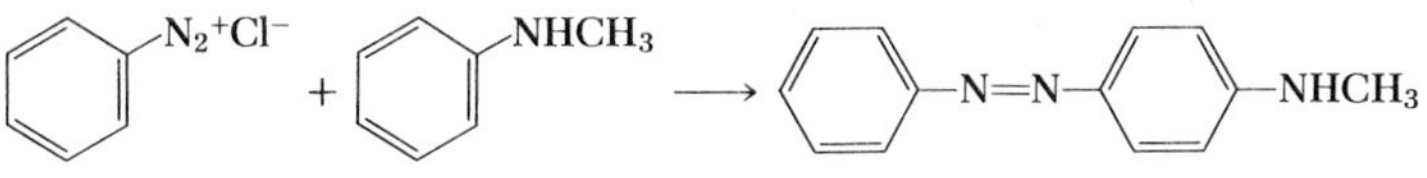

4-Methylaminoazobenzene

The azo compounds are all colored materials and many of the more complex ones, such as naphthol blue-black B, are used as dyes. Butter yellow (4-dimethylaminoazobenzene) was used for many years as an artificial coloring agent in butter and in edible oils The compound was found to be carcinogenic, however, and it is no longer used in foods.

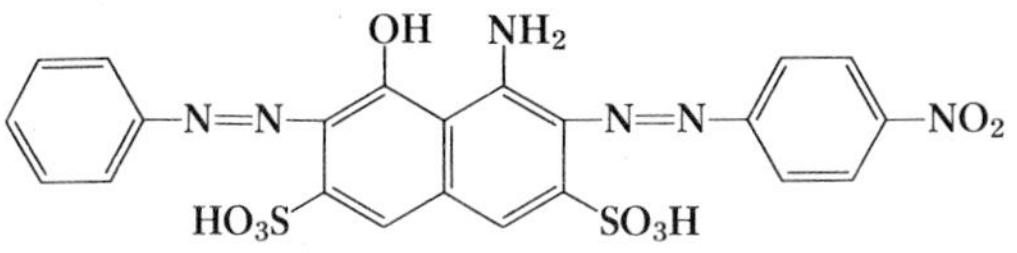

Napthol blue-black B

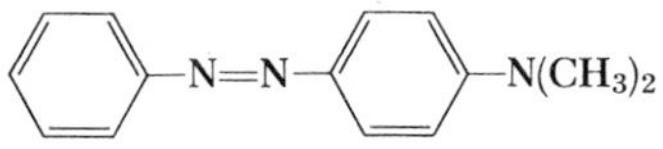

Butter yellow

Diazonium salts have been used in an ingenious way to locate in tissues and organs the sites of certain enzymes that catalyze the hydrolysis of esters. The tissue or organ is flooded with a mixture of a diazonium salt and a phenyl ester:

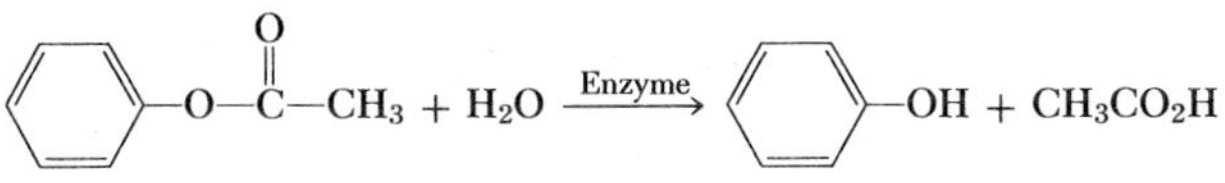

The phenol liberated reacts immediately with the diazonium salt:

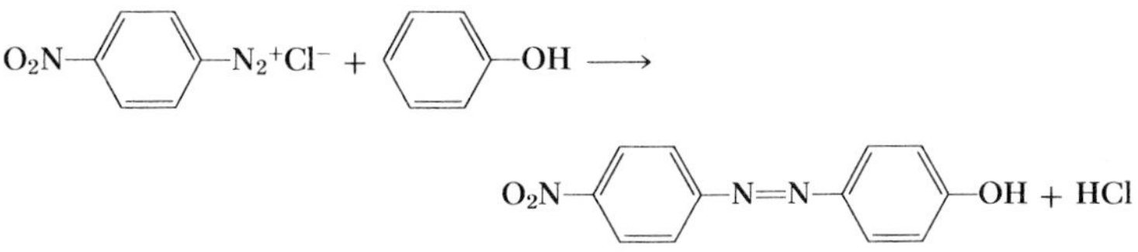

ORGANIC COMPOUNDS CONTAINING
NITROGEN, SULFUR, AND PHOSPHORUS

yielding a dye (it should be noted that the phenol esters themselves do not react with diazonium ions). The tissues are examined and the colored areas are noted; these areas contained the hydrolytic enzyme.

Diazonium salts, for example, *a* (below), react with proteins that contain tyrosine to give azo compounds with the part structure *b*.

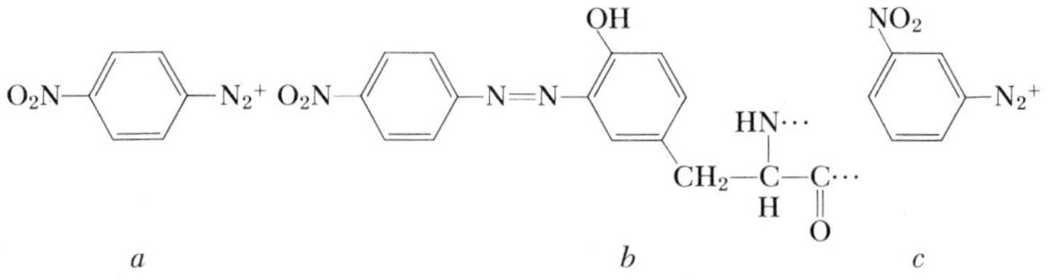

$$O_2N \text{—} \langle \rangle \text{—} N_2^+ \quad\quad O_2N \text{—} \langle \rangle \text{—} N{=}N \text{—} \langle \overset{OH}{\rangle} \text{—} \overset{\displaystyle HN\cdots}{\underset{\displaystyle CH_2{-}\overset{\textstyle }{\underset{\textstyle H}{C}}{-}\overset{\textstyle }{\underset{\textstyle O}{C}}\cdots}{}} \quad\quad O_2N \overset{NO_2}{\langle\rangle} \text{—} N_2^+$$

a	*b*	*c*

These have been injected into the bloodstream of rabbits, where they elicit the formation of antibodies directed toward the 4-nitrophenyl part of *b*. *Antibodies* are large protein molecules that form complexes with foreign species (antigens) in the blood as a first step to their removal. The antibodies are amazingly specific; those directed to *b*, for example, do not interact with antigens prepared from the 3-nitro isomer *c*, despite the fact that in such a large molecule as *b*, the difference in shape between a 4-nitrophenyl antigen (*b*) and a 3-nitrophenyl antigen (from *c*) is small indeed. No chemical method would so unerringly distinguish between the two.

THE REACTIONS OF ALIPHATIC AMINES WITH NITROUS ACID   The reactions of primary aliphatic and aromatic amines with nitrous acid are similar in that diazonium salts are formed as reaction intermediates in both cases. The subsequent reactions are quite different, however, since aliphatic diazonium salts are extremely unstable. The aliphatic diazonium ions formed as reaction intermediates decompose very rapidly to give nitrogen and a carbonium ion:

$$RNH_2 + HONO \longrightarrow RN_2^+ \longrightarrow R^+ + N_2$$

The carbonium ion then reacts in its typical fashion with any nucleophile in the system, or by the elimination of a β proton (Figure 5.6). A third typical decomposition path for carbonium ions is indicated in Figure 5.6. This is the isomerization reaction whereby carbonium ions rearrange to give a more stable carbonium ion (in this case, the *secondary* propyl carbonium ion). The reac-

Figure 5.6 The reaction of propylamine with nitrous acid.

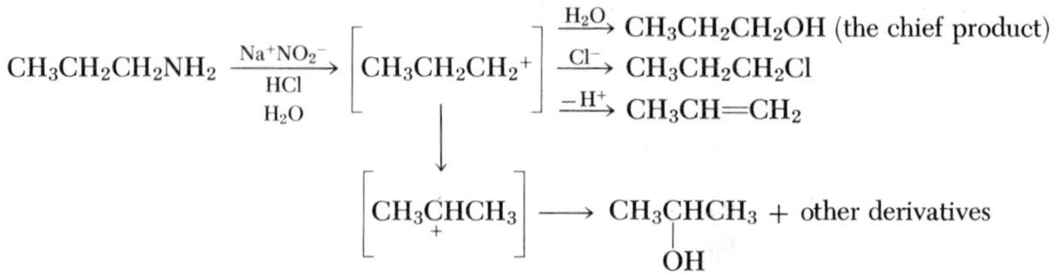

$$CH_3CH_2CH_2NH_2 \xrightarrow[\substack{HCl \\ H_2O}]{Na^+NO_2^-} \left[CH_3CH_2CH_2^+ \right] \begin{array}{l} \xrightarrow{H_2O} CH_3CH_2CH_2OH \text{ (the chief product)} \\ \xrightarrow{Cl^-} CH_3CH_2CH_2Cl \\ \xrightarrow{-H^+} CH_3CH{=}CH_2 \end{array}$$

$$\left[CH_3\overset{+}{C}HCH_3 \right] \longrightarrow \underset{OH}{CH_3\overset{|}{C}HCH_3} + \text{ other derivatives}$$

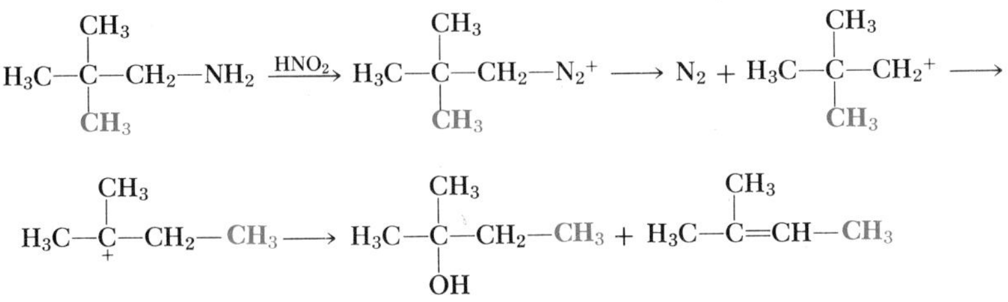

$$CH_3 \qquad\qquad CH_3 \qquad\qquad CH_3$$

$$H_3C-\underset{\underset{CH_3}{|}}{\overset{\overset{CH_3}{|}}{C}}-CH_2-NH_2 \xrightarrow{HNO_2} H_3C-\underset{\underset{CH_3}{|}}{\overset{\overset{CH_3}{|}}{C}}-CH_2-N_2^+ \longrightarrow N_2 + H_3C-\underset{\underset{CH_3}{|}}{\overset{\overset{CH_3}{|}}{C}}-CH_2^+ \longrightarrow$$

$$CH_3 \qquad\qquad CH_3 \qquad\qquad CH_3$$

$$H_3C-\underset{+}{\overset{\overset{CH_3}{|}}{C}}-CH_2-CH_3 \longrightarrow H_3C-\underset{\underset{OH}{|}}{\overset{\overset{CH_3}{|}}{C}}-CH_2-CH_3 + H_3C-\overset{\overset{CH_3}{|}}{C}=CH-CH_3$$

*Figure 5.7 **The migration of a methyl group in the reaction of 2,2-dimethyl propylamine with nitrous acid.***

tion occurs by the transfer of a hydride ion ($H:^-$) from an adjacent carbon atom:

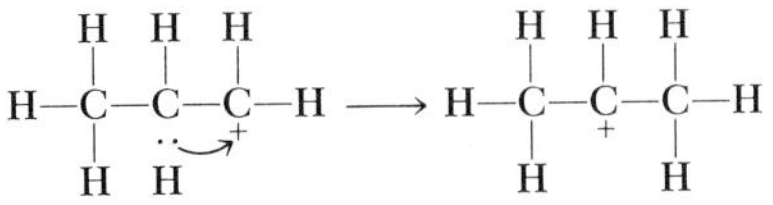

$$H-\underset{\underset{H}{|}}{\overset{\overset{H}{|}}{C}}-\underset{+}{\overset{\overset{H}{|}}{C}}-\overset{\overset{H}{|}}{\underset{\underset{H}{|}}{C}}-H \longrightarrow H-\underset{\underset{H}{|}}{\overset{\overset{H}{|}}{C}}-\underset{+}{\overset{\overset{H}{|}}{C}}-\underset{\underset{H}{|}}{\overset{\overset{H}{|}}{C}}-H$$

Alkyl groups also migrate to give more stable carbonium ions, and the tendency is so strong that the nitrous acid deamination of 2,2-dimethyl propylamine gives, as the only alcohol product, 2-methyl-2-butanol (Figure 5.7).

Migrations of alkyl groups and hydrogen of this type are typical reactions of electron-deficient species. Similar reactions are found for compounds containing electron-deficient nitrogen (Figure 5.3) and for compounds containing electron-deficient oxygen ($R-\overset{..}{O}{}^+$).

Secondary and tertiary amines react differently with nitrous acid, and the reaction serves as a convenient way to distinguish the three general types of amines. Secondary amines (both aliphatic and aromatic) react with nitrous acid to yield water-insoluble compounds called N-nitrosoamines:

$$(CH_3)_2\overset{..}{N}H + H-\overset{..}{\underset{..}{O}}-\overset{..}{N}=\overset{..}{O}: \longrightarrow (CH_3)_3\overset{..}{N}-\overset{..}{N}=\overset{..}{O}: + H_2O$$

N-Nitrosodimethylamine

Tertiary amines, on the other hand, do not react readily with nitrous acid other than to form a water-soluble salt:

$$(CH_3)_3N: + HONO \longrightarrow (CH_3)_3NH^+NO_2^-$$

Heterocyclic primary amines react with nitrous acid to give chiefly the corresponding alcohol. The reaction with cytosine (2-hydroxy-4-aminopyrimidine) is especially interesting in that cytosine is an important constituent of

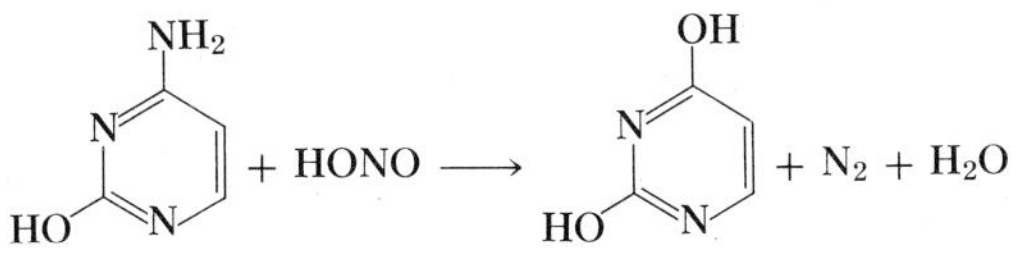

Cytosine $\qquad\qquad$ *Uracil*

DNA, the genetic material in the chromosomes of all living things. The treatment of bacterial cells with nitrous acid yields uracil in the chromosomes; this change (a mutation) causes a definite change in the progeny of the bacteria, which is passed on to all the succeeding generations!

Amides are derivatives of ammonia and the amines, in which one or more of the hydrogen atoms is replaced by an acyl group (R—C—). Examples are given in Figure 5.8. The names of the parent compounds are derived from the carboxylic acid names by the substitution of the suffix -*amide* for the -*ic acid* endings. The amides have many of the properties of the amines. They are basic compounds, but are much more weakly basic than are the amines because of resonance delocalization of the unshared electron pair of nitrogen:

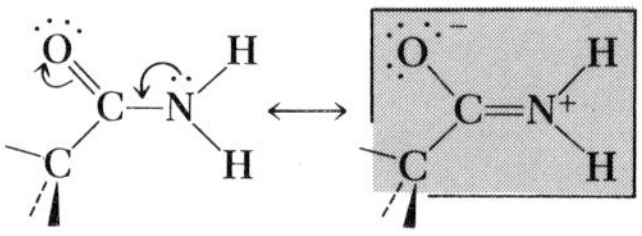

As the second resonance contributor indicates, the C—N bond of an amide has considerable double-bond character. As a result, rotation about this bond is very slow. Also, the amide grouping has the geometry of the structurally related molecule, ethylene

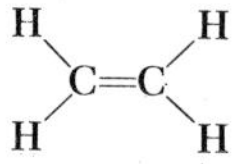

That is, all the atoms shown within the rectangle above are in one plane. The planar amide group plays a large role in the structure of peptides and proteins.

Figure 5.8   ***Amides of aliphatic and aromatic carboxylic acids.***

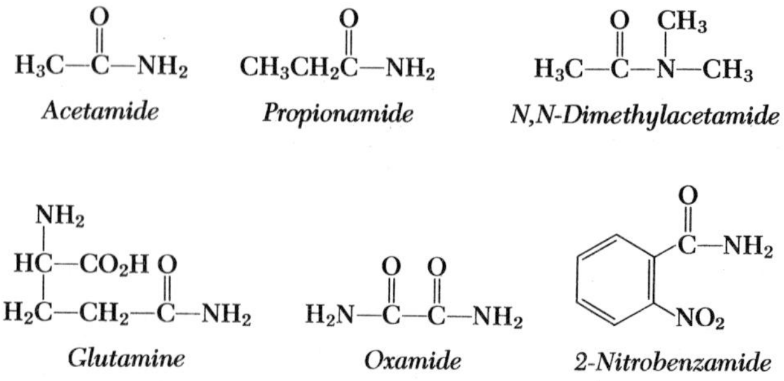

Acetamide

Propionamide

N,N-Dimethylacetamide

Glutamine

Oxamide

2-Nitrobenzamide

Amides are prepared, principally, by the reaction of ammonia (or an amine) with an ester or an acid anhydride:

$$\underset{\text{Methyl benzoate}}{\text{C}_6\text{H}_5\overset{\text{O}}{\underset{\|}{\text{C}}}\text{—O—CH}_3 + :\text{NH}_3} \longrightarrow \left[\text{C}_6\text{H}_5\underset{\text{NH}_2}{\overset{\text{OH}}{\underset{|}{\overset{|}{\text{C}}}}}\text{—O—CH}_3 \right] \longrightarrow$$

$$\underset{\text{Benzamide}}{\text{C}_6\text{H}_5\overset{\text{O}}{\underset{\|}{\text{C}}}\text{—NH}_2 + \text{CH}_3\text{OH}}$$

$$\underset{\text{Acetyl chloride}}{\text{CH}_3\overset{\text{O}}{\underset{\|}{\text{C}}}\text{—Cl} + 2\,(\text{CH}_3\text{CH}_2)_2\overset{..}{\text{N}}\text{H}} \longrightarrow \underset{\text{N,N-Diethylacetamide}}{\text{CH}_3\overset{\text{O}}{\underset{\|}{\text{C}}}\text{—N(CH}_2\text{CH}_3)_2} + (\text{CH}_3\text{CH}_2)_2\text{NH}_2{}^+\text{Cl}^-$$

REACTIONS OF THE AMIDES Amides are readily hydrolyzed in both acidic and basic solutions:

(1) *Acid hydrolysis:* $\text{R}—\overset{\text{O}}{\underset{\|}{\text{C}}}—\text{NH}_2 + \text{H}_3\text{O}^+\text{Cl}^- \longrightarrow \text{R}—\overset{\text{O}}{\underset{\|}{\text{C}}}—\overset{+}{\text{N}}\text{H}_3 \;\; \text{Cl}^- + \text{H}_2\text{O} \longrightarrow$

$$\text{R}—\underset{\underset{\text{Cl}^-}{\overset{|}{\overset{+}{\text{O}}\text{H}_2}}}{\overset{\text{O}^-}{\overset{|}{\underset{|}{\text{C}}}}}—\overset{+}{\text{N}}\text{H}_3 \longrightarrow \text{R}—\underset{\text{Cl}^-}{\overset{\text{O}}{\underset{\|}{\text{C}}}}—\overset{+}{\text{O}}\text{H}_2 + \text{NH}_3 \longrightarrow \text{RCO}_2\text{H} + \text{NH}_4{}^+\text{Cl}^-$$

(2) *Base hydrolysis:* $\text{R}—\overset{\text{O}}{\underset{\|}{\text{C}}}—\text{N(CH}_3)_2 + \text{Na}^+\text{OH}^- \longrightarrow \text{R}—\underset{\text{OH}}{\overset{\text{O}^-\text{Na}^+}{\underset{|}{\overset{|}{\text{C}}}}}—\text{N(CH}_3)_2 \longrightarrow$

$$\text{R}—\overset{\text{O}}{\underset{\|}{\text{C}}}—\text{OH} + (\text{CH}_3)_2\text{N}^-\text{Na}^+ \longrightarrow \text{RCO}_2{}^-\text{Na}^+ + (\text{CH}_3)_2\text{NH}$$

This reaction is often the first step in the identification of amides since the carboxylic acid and amine obtained can usually be identified more easily than the amide itself. A second general reaction of amides, the Hofmann rearrangement, was outlined in the section on the preparation of amines.

Dehydration of amides Most unsubstituted amides are dehydrated by many reagents that react with water, such as phosphorus pentoxide, P_2O_5, and benzenesulfonyl chloride,

$$\text{C}_6\text{H}_5\text{—SO}_2\text{Cl}$$

the reaction mechanisms are presumably similar:

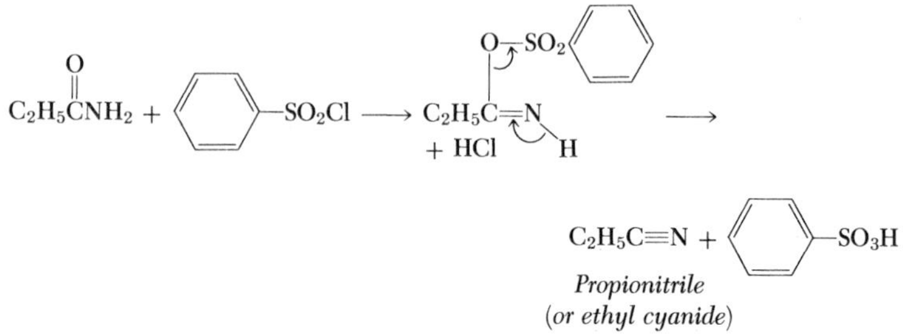

$$C_2H_5\overset{\overset{\displaystyle O}{\|}}{C}NH_2 + \langle \rangle-SO_2Cl \longrightarrow C_2H_5C{=}N + HCl \qquad$$

$$C_2H_5C{\equiv}N + \langle \rangle-SO_3H$$

Propionitrile
(*or ethyl cyanide*)

An alternative method of synthesizing this compound was given earlier (on p. 97).

The *amino acids,*

$$R-\overset{\overset{\displaystyle H}{|}}{\underset{\underset{\displaystyle NH_2}{|}}{C}}-CO_2H$$

are carboxylic acids substituted on the α-carbon atom by an amino group. They were originally obtained by hydrolysis of proteins, but by the early part of the twentieth century they had all been synthesized in the laboratory. Glycine (aminoacetic acid) is the simplest amino acid, and it was also the first to have been isolated; it was obtained in crystalline form in 1820 by Bracconnot from the hydrolysis products of gelatin. Scientists have isolated a few dozen amino acids from natural sources since that time, largely from proteins. The 20 amino acids that are essential for the synthesis of proteins in living systems are listed in Table 5.2. Note the similarities in what are essentially the building blocks of the proteins.

Since carboxylic acids are neutralized by amines, it is not surprising that amino acids, which contain both of these groups, exist as internal salts (*dipolar ions*). The form shown is the principal one for amino acids at pH 7 (essentially that of cellular fluids).

$$R-\overset{\overset{\displaystyle H}{|}}{\underset{\underset{\displaystyle H-\overset{+}{N}-H}{|}}{C}}-\overset{\overset{\displaystyle O}{\|}}{C}-O^-$$

When an amino acid is treated with a base, the proton is lost from the nitrogen atom of the $-NH_3^+$ group:

Table 5.2 Amino acids:

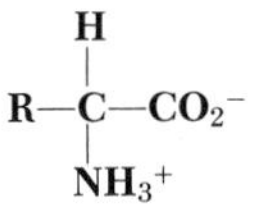

$$R-\underset{\underset{NH_3^+}{|}}{\overset{\overset{H}{|}}{C}}-CO_2^-$$

R	NAME	R	NAME
H—	Glycine	HO—⟨benzene⟩—CH$_2$—	Tyrosine
CH$_3$—	Alanine		
HO—CH$_2$—	Serine	⟨imidazole ring, HN, N⟩—CH$_2$—	Histidine
HS—CH$_2$—	Cysteine		
CH$_3$CHOH—	Threonine	⟨indole ring, N-H⟩—CH$_2$—	Tryptophane
HO$_2$C—CH$_2$—	Aspartic acid		
(CH$_3$)$_2$CH—	Valine		
H$_3$C—S—CH$_2$CH$_2$—	Methionine		
HO$_2$C—CH$_2$CH$_2$—	Glutamic acid		
(CH$_3$)$_2$CHCH$_2$—	Leucine		
CH$_3$CH$_2$CH(CH$_3$)—	Isoleucine		
H$_2$N—CH$_2$CH$_2$CH$_2$CH$_2$—	Lysine		

Two cyclic amino acids

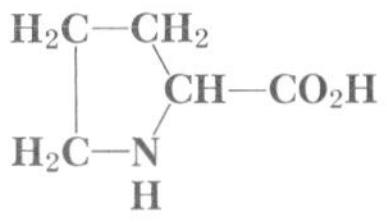

$$H_2N-\overset{\overset{NH}{\|}}{C}-NH-CH_2-CH_2-CH_2- \qquad Arginine$$

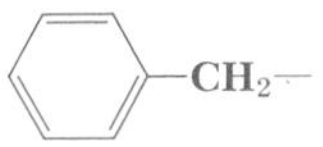

$$HO_2C-\underset{\underset{NH_2}{|}}{CH}-CH_2-S-S-CH_2- \qquad Cystine$$

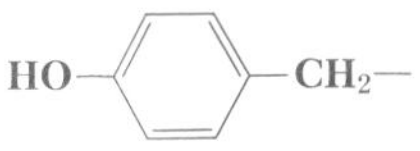

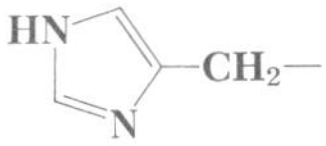

structure	name
⟨proline ring: H$_2$C—CH$_2$ / CH—CO$_2$H / H$_2$C—N—H⟩	Proline
⟨hydroxyproline ring: OH, HC—CH$_2$ / CH—CO$_2$H / H$_2$C—N—H⟩	Hydroxyproline

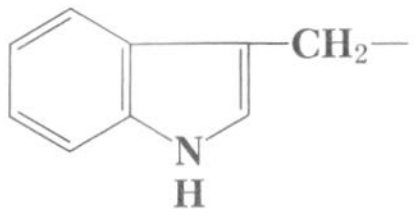

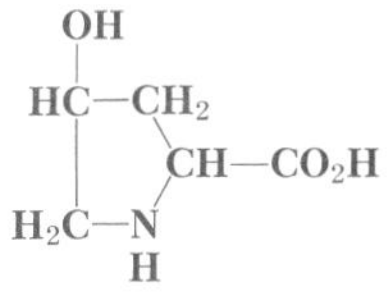

⟨phenyl⟩—CH$_2$—   Phenylalanine

$$H-\underset{\underset{\overset{|}{H}}{\overset{|}{N^+}-H}}{\overset{\overset{H}{|}}{C}}-\overset{\overset{O}{\|}}{C}-O^- + H_2O \rightleftharpoons H_3O^+ + H-\underset{\underset{H}{|}}{\overset{\overset{H}{|}}{C}}-\overset{\overset{O}{\|}}{C}-O^-\ \ (:\!N-H)$$

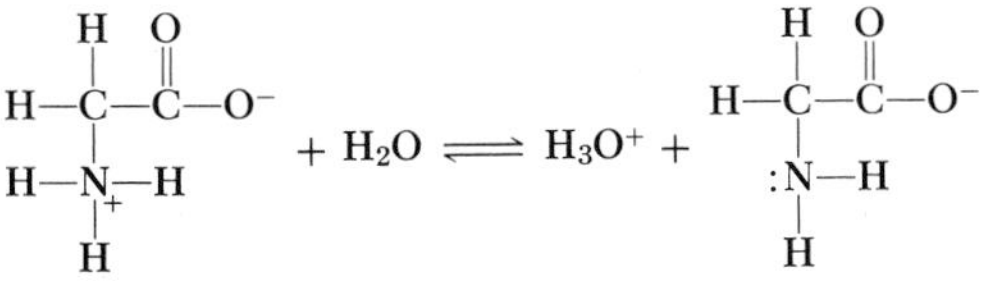

As a result, the ionization constants are different from those of the carboxylic acids, in which the proton is transferred from an oxygen atom to water. The ionization constant for glycine is 1.6×10^{-10} compared to the value of 1.8×10^{-5} found for acetic acid. In strong acids, the amino acids behave as bases, and a proton is transferred to the carboxylate ion:

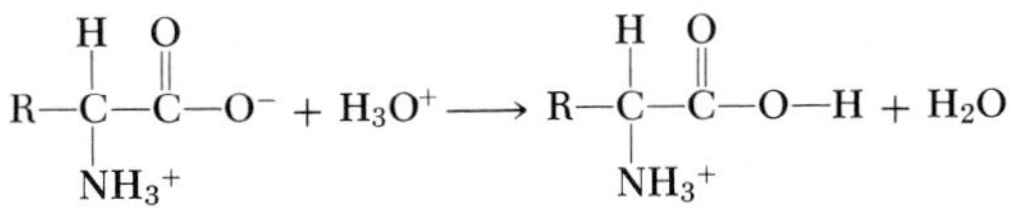

$$R-\underset{\underset{NH_3^+}{|}}{\overset{\overset{H}{|}}{C}}-\overset{\overset{O}{\|}}{C}-O^- + H_3O^+ \longrightarrow R-\underset{\underset{NH_3^+}{|}}{\overset{\overset{H}{|}}{C}}-\overset{\overset{O}{\|}}{C}-O-H + H_2O$$

The amino acids are thus typical amphoteric compounds in that they react with both acids and bases.

PREPARATION OF THE AMINO ACIDS   Most of the amino acids may be prepared by the displacement of a bromide ion from an α-bromo acid with ammonia:

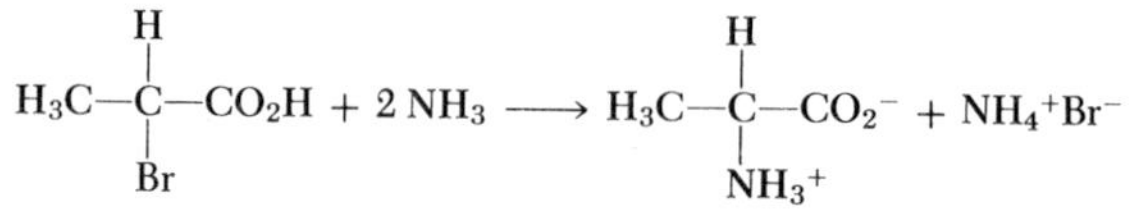

$$\underset{\underset{Br}{|}}{\overset{\overset{H}{|}}{H_3C-C-CO_2H}} + 2\,NH_3 \longrightarrow \underset{\underset{NH_3^+}{|}}{\overset{\overset{H}{|}}{H_3C-C-CO_2^-}} + NH_4^+Br^-$$

Alanine

A second general method, called the *Strecker synthesis,* involves the reaction of an aldehyde with ammonium cyanide and the hydrolysis of the nitrile formed as an intermediate.

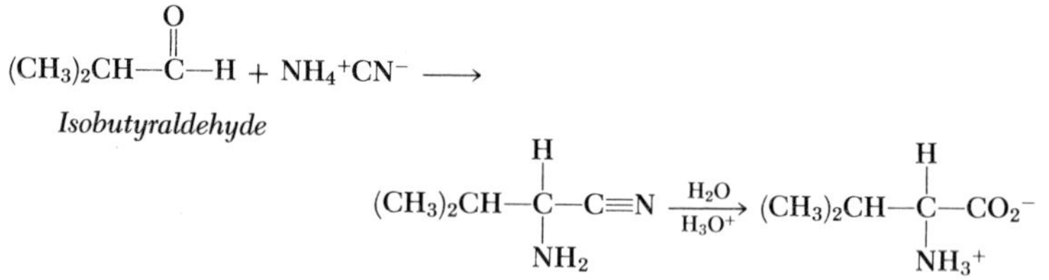

$$(CH_3)_2CH-\overset{\overset{\textstyle O}{\|}}{C}-H + NH_4^+CN^- \longrightarrow$$

Isobutyraldehyde

$$\underset{\underset{NH_2}{|}}{\overset{\overset{H}{|}}{(CH_3)_2CH-C-C{\equiv}N}} \xrightarrow[H_3O^+]{H_2O} \underset{\underset{NH_3^+}{|}}{\overset{\overset{H}{|}}{(CH_3)_2CH-C-CO_2^-}}$$

Valine

Many other synthetic methods have been devised, which are specific for one or more of the amino acids; for further information, consult the list of selected readings at the end of this volume.

A number of different biological syntheses of the amino acids are known, but a more or less general synthesis involves α-keto acids and pyridoxamine phosphate (symbolized $R'CH_2NH_2$):

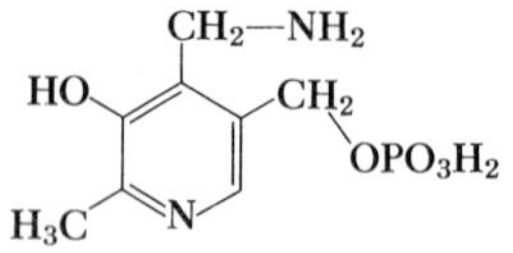

Pyridoxamine phosphate

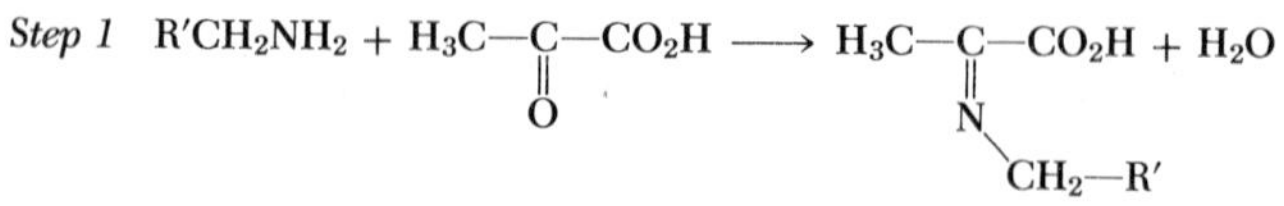

Step 1 $\quad R'CH_2NH_2 + \underset{\underset{O}{\|}}{H_3C-C-CO_2H} \longrightarrow \underset{\underset{N}{\diagdown}\atop CH_2-R'}{H_3C-C-CO_2H} + H_2O$

Pyruvic acid $\qquad\qquad$ X

Step 2 $\quad X \longrightarrow \underset{\underset{N=C-R'}{|}\atop\underset{H}{|}}{\overset{\overset{H}{|}}{H_3C-C-CO_2H}}$

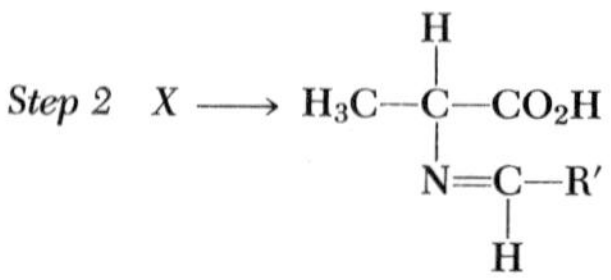

Y

$$\text{Step 3} \quad Y + H_2O \longrightarrow \underset{\underset{NH_2}{|}}{\overset{\overset{H}{|}}{CH_3CCO_2H}} + R'CHO$$

*Alanine   Pyridoxal
phosphate*

Step 1 is a typical carbonyl-amine condensation (Figure 4.8), step 2 a tautomerization (p. 104), and step 3 a typical hydrolysis—a reaction with a mechanism that is essentially the reverse of that of step 1. The by-product in this synthesis of amino acids is pyridoxal phosphate. We have already seen that this compound is a critical reactant in decarboxylation (p. 123). In addition, pyridoxal phosphate is involved in the oxidation of amino acids, in the racemization of amino acids, and in the modification of the side chain of amino acids. Although we cannot consider the mechanisms at this point, they are all related to the present mechanism in involving condensation products of the type Y.

REACTIONS OF THE AMINO ACIDS   The reactions of amino acids are similar to those of the amines and carboxylic acids:

$$H_3C\underset{\underset{NH_3^+}{|}}{-CH}-CO_2^- \xrightarrow[H_3O^+]{CH_3OH} H_3C\underset{\underset{NH_2}{|}}{-CH}-\overset{\overset{O}{\|}}{C}-O-CH_3 \xrightarrow{CH_3CCl} H_3C\underset{\underset{HN-C-CH_3}{\underset{\underset{O}{\|}}{|}}}{-CH}-CO_2CH_3$$

An especially important reaction of amino acids and their derivatives is amide formation, as we shall illustrate further in the section of proteins (Chapter 6).

$$H_3C\underset{\underset{\underset{O}{\|}}{\underset{H-N-C-CH_3}{|}}}{-CH}-CO_2CH_3 + CH_3NH_2 \longrightarrow H_3C\underset{\underset{\underset{O}{\|}}{\underset{H-N-C-CH_3}{|}}}{-CH}-\overset{\overset{O}{\|}}{C}-NH\,CH_3 + CH_3OH$$

Two amino acids can be linked by an amide bond to give compounds called dipeptides, as shown below. All of the possible dipeptides, as well as a good many

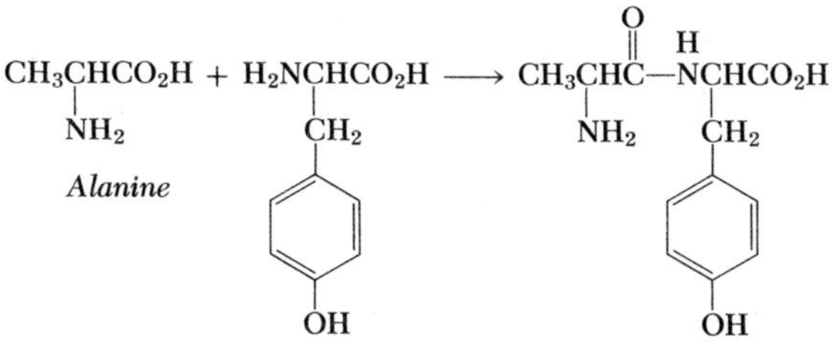

$$CH_3\underset{\underset{NH_2}{|}}{CH}CO_2H + H_2N\underset{\underset{CH_2}{|}}{CH}CO_2H \longrightarrow CH_3\underset{\underset{NH_2}{|}}{CH}\overset{\overset{O}{\|}}{C}-\overset{\overset{H}{}}{N}\underset{\underset{CH_2}{|}}{CH}CO_2H$$

Alanine   *Tyrosine   Alanyltyrosine*

of tri-, tetra-, and pentapeptides, and so on, have been synthesized in the laboratory. The technique usually involves making chemical derivatives (so-called blocked derivatives) of all the amino groups in one molecule and all the carboxylic groups in the other. An amide bond is then formed between the free amino group of the latter molecule and the carboxyl group of the former (usually in an activated form). At this point, all the blocking groups are removed, yielding the dipeptide. The biological syntheses of peptides and proteins will be discussed in Chapter 6.

The essential amino acids cysteine and methionine, the antibiotic penicillin, the vitamin thiamine, coenzyme A (p. 157), glutathione (p. 97), firefly luciferin, the mercapturic acids (p. 92), and lipoic acid (p. 149) are all compounds of biological importance that contain sulfur. In these compounds sulfur is in a -2 oxidation stage and the chemistry is related to that of oxygen, as may be expected from the positions of the two elements in the periodic table. The chief differences are the greater acidity of the SH bond relative to that of the OH bond and the greater nucleophilicity of the sulfur compounds. Also, sulfur has a larger number of valence states than does oxygen, and compounds with a positive oxidation number, such as the sulfonic acids (RSO_3H), are of considerable importance in chemistry.

Examples of some simple sulfur-containing compounds are given in Figure 5.9. The names of these compounds are closely related to the names of the oxygen analogs; usually the term *thio-* is added to indicate the presence of sulfur (the RSH compounds are very often called *mercaptans,* however). The thiols are characterized by atrocious odors (butanethiol is chiefly responsible for the scent of the skunk) and by the formation of insoluble precipates with lead, mercury, and the other heavy metals. Many enzymes contain thiol groups, and it is common practice to precipitate or inactivate these enzymes with mercuric salts.

The thiols are generally prepared by displacement reactions involving alkyl halides and salts of hydrogen sufide (H_2S):

$$CH_3CH_2-Br + K^+SH^- \longrightarrow CH_3CH_2SH + K^+Br^-$$

Figure 5.9   **Compounds of sulfur.**

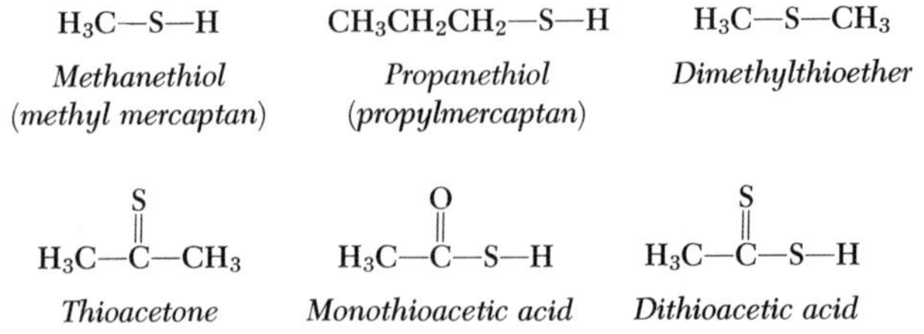

H_3C-S-H	$CH_3CH_2CH_2-S-H$	$H_3C-S-CH_3$
Methanethiol (methyl mercaptan)	*Propanethiol* (propylmercaptan)	*Dimethylthioether*

$H_3C-\overset{\overset{S}{\|}}{C}-CH_3$	$H_3C-\overset{\overset{O}{\|}}{C}-S-H$	$H_3C-\overset{\overset{S}{\|}}{C}-S-H$
Thioacetone	*Monothioacetic acid*	*Dithioacetic acid*

The thioethers are prepared in a similar way:

$$CH_3CH_2-Br + CH_3S^-K^+ \longrightarrow CH_3SCH_2CH_3 + K^+Br^-$$

Methylethylthioether

Thiols are more acidic than are the corresponding alcohols, although they are still weak acids; the ionization constant for ethanethiol is 10^{-12}, for example:

$$CH_3CH_2SH + H_2O \rightleftharpoons CH_3CH_2S^- + H_3O^+ \qquad K = 10^{-12}$$

The negative ions of thiols can be prepared quantitatively by the reaction of the thiol with sodium hydroxide; these ions are readily oxidized by the oxygen in air or by oxidizing agents, such as I_2, to give the corresponding free radicals:

$$2\,H_3C-\overset{..}{\underset{..}{S}}{:}^-Na^+ + \,:\!\overset{..}{\underset{..}{I}}\!:\!\overset{..}{\underset{..}{I}}\!: \longrightarrow 2\,H_3C-\overset{..}{\underset{..}{S}}\cdot + 2\,Na^+:\!\overset{..}{\underset{..}{I}}\!:$$

The radicals are short-lived intermediates and they stabilize themselves through dimerization to yield compounds known as, *disulfides.*

$$2\,H_3C-\overset{..}{\underset{..}{S}}\cdot \longrightarrow H_3C-\overset{..}{\underset{..}{S}}-\overset{..}{\underset{..}{S}}-CH_3$$

Dimethyl disulfide

Protein chains that contain SH groups (largely from the amino acid cysteine in the protein) also undergo oxidation to form S—S bonds. As we shall see in Chapter 6, this is a general method for linking up two separate protein chains or for constructing a new ring out of a linear molecule (Figure 6.1).

At least one disulfide, lipoic acid, is of considerable importance in cellular metabolism. This acid reacts as an oxidizing agent in certain cellular reactions, such as the oxidative decarboxylation of pyruvic acid (CH_3COCO_2H), and in these reactions, the lipoic acid is reduced to the corresponding dithiol; oxidation at a later stage by other components of the cell then regenerates the lipoic acid.[*]

$$\underset{\overset{|}{S}\rule{1.2cm}{0.4pt}\overset{}{S}}{CH_2CH_2CHCH_2CH_2CH_2CH_2CO_2H} \qquad\qquad \underset{\overset{|}{SH}\quad\overset{|}{SH}}{CH_2CH_2CHCH_2CH_2CH_2CH_2CO_2H}$$

Lipoic acid   *Reduced form of lipoic acid*

The oxidation of thioethers with hydrogen peroxide yields monoxides called *sulfoxides,* and these in turn can be oxidized to dioxides called *sulfones:*

$$H_3C-\overset{..}{\underset{..}{S}}-CH_3 \xrightarrow{H_2O_2} H_3C-\overset{\overset{\displaystyle :\overset{..}{O}:}{|}}{\underset{..}{S}}-CH_3 \xrightarrow{H_2O_2} H_3C-\overset{\overset{\displaystyle \overset{..}{O}:}{|}}{\underset{\underset{\displaystyle :\overset{..}{O}:}{|}}{S}}-CH_3$$

Dimethyl sulfoxide   *Dimethyl sulfone*

[*] For details of the biological reactions of lipoic acid, see W. D. McElroy, *Cell Physiology and Biochemistry,* 2nd ed. (Englewood Cliffs, N.J.: Prentice-Hall, Inc., 1964).

Dimethyl sulfoxide is a highly polar solvent capable of dissolving both organic and inorganic compounds. It has shown promise as a solvent for proteins, and recently it was used in place of water as a solvent in a study of the enzyme trypsin.

THE SULFONIC ACIDS AND DERIVATIVES   The oxidation of thiols under acidic conditions leads to a series of acids, the most important of which are the sulfonic acids:

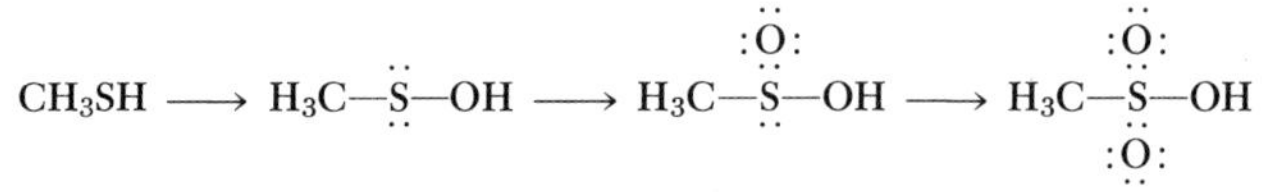

$$CH_3SH \longrightarrow H_3C-\overset{..}{\underset{..}{S}}-OH \longrightarrow H_3C-\overset{:\overset{..}{O}:}{\underset{..}{S}}-OH \longrightarrow H_3C-\overset{:\overset{..}{O}:}{\underset{:\underset{..}{O}:}{S}}-OH$$

Methane   *Methane*   *Methane*
sulfenic acid   *sulfinic acid*   *sulfonic acid*

A similar sequence is known for the amino acid cysteine, the final product—cysteic acid,

$$HO_2C-CH(NH_2)-CH_2-SO_3H$$

being an intermediate in the biological breakdown of cysteine. The aromatic sulfonic acids are usually prepared directly, on the other hand, by the action of sulfuric acid on the aromatic compounds:

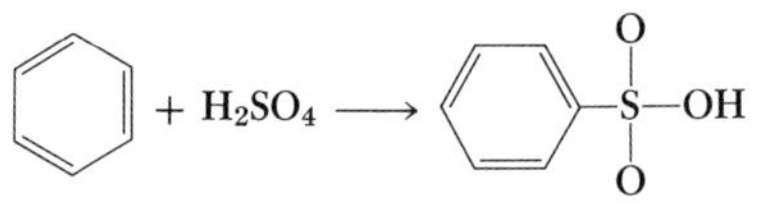

Benzenesulfonic acid

The synthesis of an interesting derivative of an aromatic sulfonic acid is given in Figure 5.10. The derivative, called *saccharin,* is widely used as an artificial sweetening agent.

The sulfonic acids are related to sulfuric acid, and like sulfuric acid they are strong acids, becoming fully ionized in water solutions:

*Figure 5.10   **The   synthesis   of saccharin.***

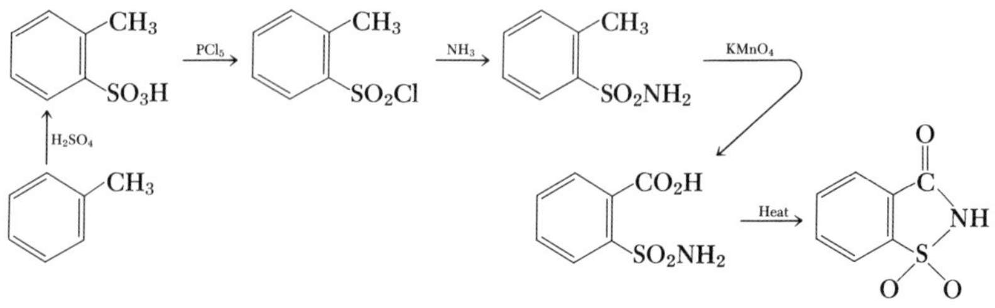

$$RSO_3H + H_2O \longrightarrow RSO_3^- + H_3O^+$$

The sulfonic acids show many of the chemical properties of the carboxylic acids; they are readily converted into the corresponding acid chlorides and esters, for example:

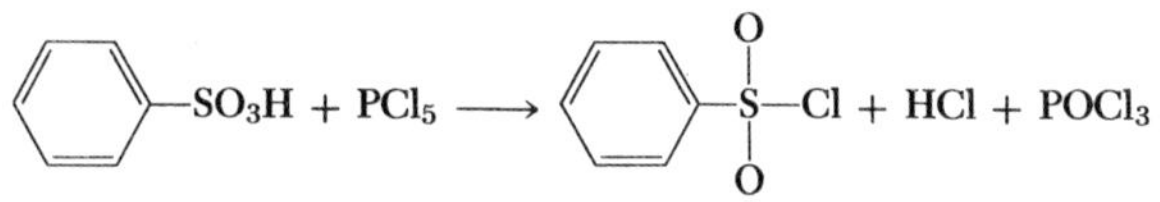

Benzenesulfonyl chloride

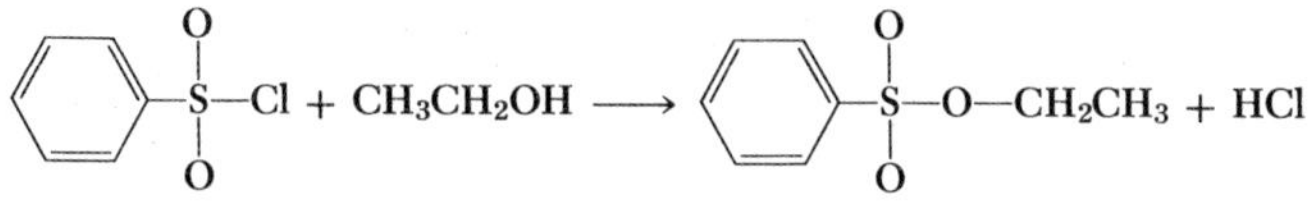

Ethyl benzenesulfonate

Esters of sulfuric acid itself may also be prepared by reactions of this type:

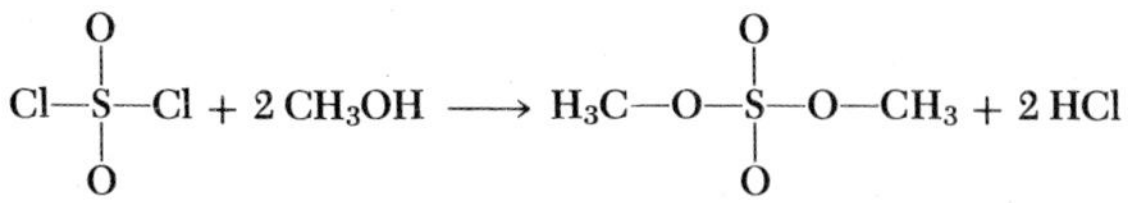

Sulfuryl chloride   *Dimethyl sulfate*

Esters of sulfuric acid, especially those involving the hydroxyl group of a sugar molecule, are of considerable biological importance. The anticlotting agent, heparin, for example, is a polysaccharide (p. 179) which contains such an ester linkage to sulfuric acid.

NUCLEOPHILIC DISPLACEMENT REACTIONS OF COMPOUNDS CONTAINING SULFUR   The alkyl sulfates and sulfonates react as alkylating agents with nucleophiles:

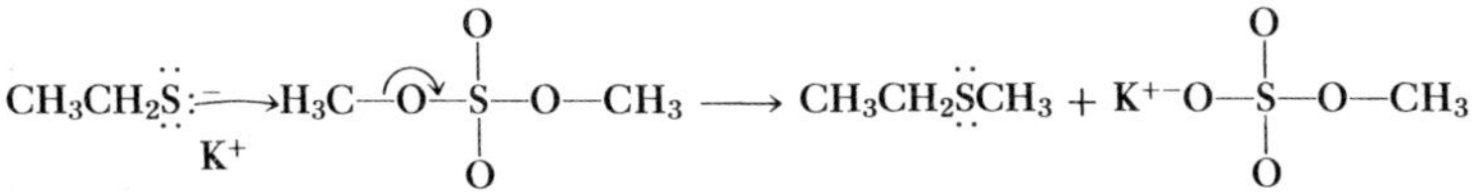

and as such can be used interchangeably with the alkyl halides (CH_3I, for example). A closely related alkylation reaction occurs with the thioethers to yield salts that are similar in structure to the hydronium and quaternary ammonium ions.

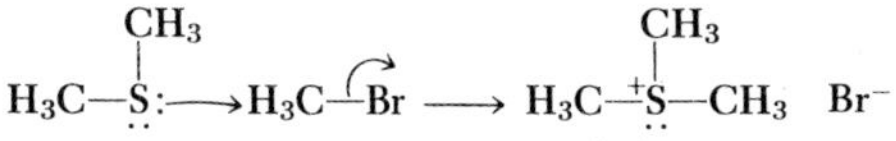

Trimethylsulfonium bromide

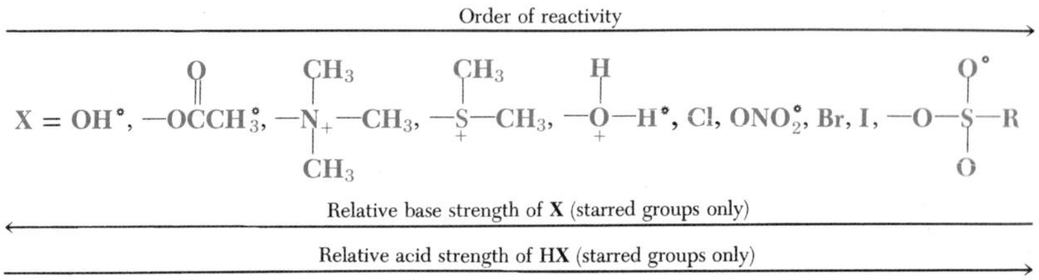

$$X = OH^\circ, \quad -OCCH_3^\circ, \quad -\overset{CH_3}{\underset{CH_3}{N_+}}-CH_3, \quad -\overset{CH_3}{\underset{+}{S}}-CH_3, \quad -\overset{H}{\underset{+}{O}}-H^\circ, \quad Cl, \quad ONO_2^\circ, \quad Br, \quad I, \quad -O-\overset{O^\circ}{\underset{O}{S}}-R$$

Relative base strength of **X** (starred groups only)

Relative acid strength of **HX** (starred groups only)

Figure 5.11 **The reactivity of various methylating agents, H_3C—X.**

Compounds of this type are also alkylating agents:

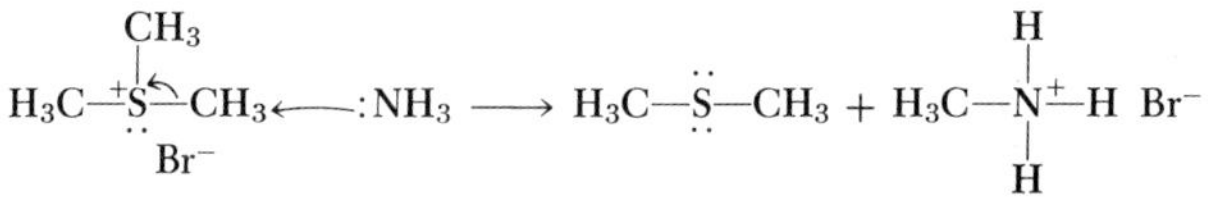

$$H_3C-\overset{CH_3}{\underset{Br^-}{\overset{+}{S}}}-CH_3 \longleftarrow :NH_3 \longrightarrow H_3C-\ddot{S}-CH_3 + H_3C-\overset{H}{\underset{H}{\overset{+}{N}}}-H \ \ Br^-$$

The sulfonium salts constitute the third group of alkylating agents that have been discussed. A general reaction can be written for these reactions,

$$H_3N: \longrightarrow H_3C-X^{+ \text{ or } 0} \longrightarrow CH_3NH_3^+ \ X^{0 \text{ or } -}$$

where $X = I^-$, $R-SO_3^-$, or $R-\ddot{S}-R$. The group displaced, X, is generally called the leaving group, and in the examples cited, it is either a neutral species or the negative ion of a strong acid (I^- or RSO_3^-, for example). These species are all weak bases and, in fact, the correlation of the basicity of the leaving group (which is, in turn, related to the stability of the groups) with the rate of the displacement reaction generally is good. Alkyl nitrates ($H_3C-O-NO_2$) and quaternary salts ($H_3C-NR_3^+$) are also good alkylating agents, for example. Similar arguments account for the fact that methanol is a very poor alkylating agent (methanol does not react with sodium iodide), whereas the protonated form is a rather good alkylating agent.

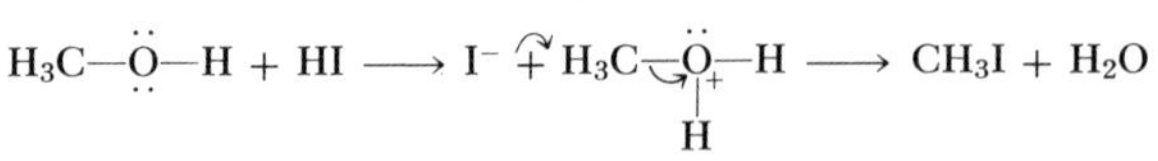

$$H_3C-\ddot{O}-H + HI \longrightarrow I^- + H_3C-\overset{}{\underset{H}{\overset{+}{O}}}-H \longrightarrow CH_3I + H_2O$$

It is thus possible to set up a series of "leaving groups" in order of the rates at which they are displaced from their methyl derivatives by a nucleophile (Figure 5.11).

As long as we examine leaving groups that are attached to the methyl by the same atom (oxygen in the case of the starred groups, Figure 5.11), then a relationship similar to that outlined on p. 155 holds: the weaker the base displaced, the faster the reaction (that is, the better the leaving group).

CHEMICAL BACKGROUND FOR THE
BIOLOGICAL SCIENCES

It is interesting to note that certain compounds necessary in the diet of animals are methyl-transfer agents (that is, alkylating agents), and that the methyl donating group is similar to those given in Figure 5.11. One compound of this type, choline, was discussed on page 134; other examples are given below:

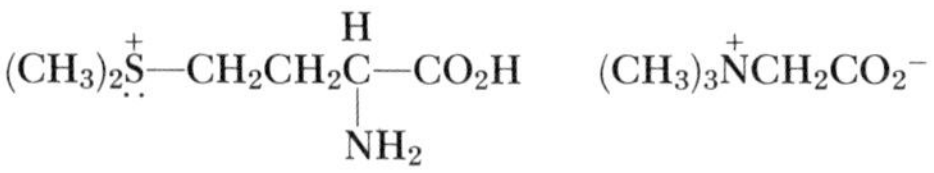

$$(CH_3)_2\overset{+}{\underset{..}{S}}-CH_2CH_2\overset{\overset{\displaystyle H}{|}}{C}-CO_2H \qquad (CH_3)_3\overset{+}{N}CH_2CO_2^-$$
$$\underset{NH_2}{|}$$

Methyl sulfonium derivative   *Betaine*
of methionine

The mechanism of alkylation in a cell is almost certainly related to the mechanism of the alkylations carried out in the test tube; that is, nucleophilic substitution is involved in both cases.

A great variety of phosphorus compounds are known. In this volume, we shall mention only those with oxidation numbers of -3 and $+5$.

The phosphorus analog of ammonia is called phosphine, PH_3, and the derivatives of this compound are named largely as derivatives of phosphine (Figure 5.12). Ammonia and phosphine are similar in the types of reactions they undergo; they differ qualitatively, however, in that phosphine has the greater acidity and nucleophilicity—a result of the lower position of phosphorus in the periodic table. Most of the alkyl derivatives of phosphine are prepared from the alkyl halides; typical nucleophilic displacements are involved:

$$PH_3 + 3\ CH_3I \xrightarrow{\ 3\ KOH\ } (CH_3)_3P + 3\ K^+I^- + 3\ H_2O$$

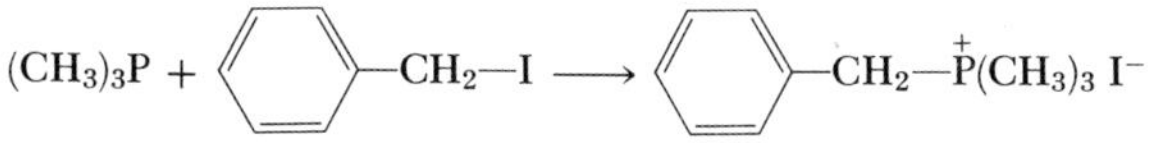

$$(CH_3)_3P + \langle\!\!\bigcirc\!\!\rangle\!-CH_2-I \longrightarrow \langle\!\!\bigcirc\!\!\rangle\!-CH_2-\overset{+}{P}(CH_3)_3\ I^-$$

Benzyltrimethylphosphonium iodide

Figure 5.12   ***Organic compounds of phosphorus in the -3 stage of oxidation.***

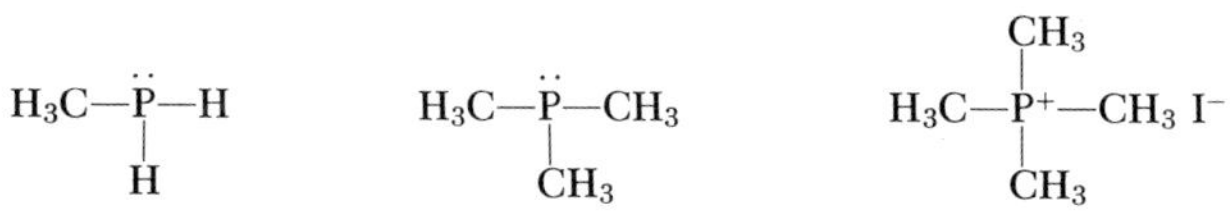

$$\underset{\overset{|}{H}}{\overset{\displaystyle ..}{H_3C-P-H}} \qquad \underset{\overset{|}{CH_3}}{\overset{\displaystyle ..}{H_3C-P-CH_3}} \qquad \underset{\overset{|}{CH_3}}{\overset{\overset{\displaystyle CH_3}{|}}{H_3C-P^+-CH_3}}\ I^-$$

Methylphosphine   *Trimethylphosphine*   *Tetramethylphosphonium*
iodide

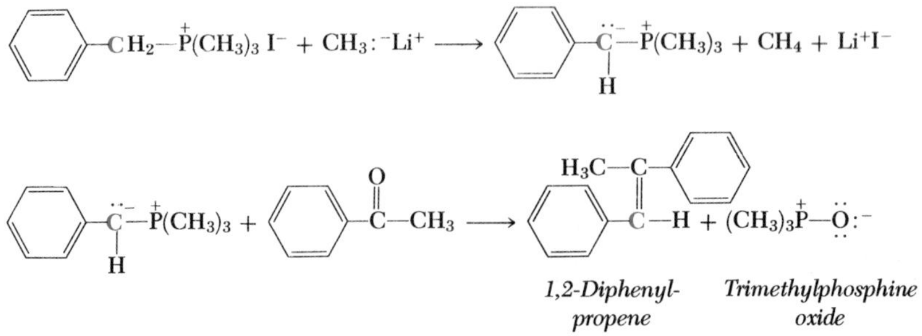

Figure 5.13 **The Wittig reaction, a synthesis of alkenes from the corresponding carbonyl compounds.**

An extremely valuable reaction of the phosphonium halides known as the *Wittig reaction* has been developed recently (Figure 5.13); in effect, this reaction is a direct method for the conversion of carbonyl compounds into alkene derivatives. In the example cited, the proton adjacent to the phenyl ring is removed by the base methyl lithium because it is the most acidic proton; the electron pair formed in this position is stabilized by resonance:

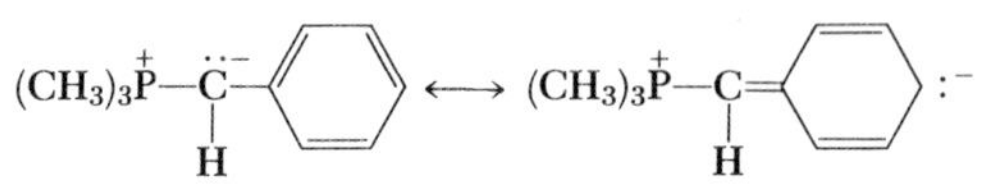

The reaction is a general one, and by the use of different phosphonium compounds, it can be used to substitute the $=CH_2$, $=CHR$, and $=CR_2$ groups for the doubly bonded oxygen atoms in most aldehydes and ketones.

DERIVATIVES OF PHOSPHORIC ACID Phosphoric acid (H_3PO_4) is representative of the fully oxidized compounds of phosphorus (oxidation number, $+5$). Alkyl esters of phosphoric acid are prepared in much the same way as are esters of sulfuric acid and the carboxylic acids. One of the simplest methods involves the reaction of a phosphoric acid chloride with an alcohol:

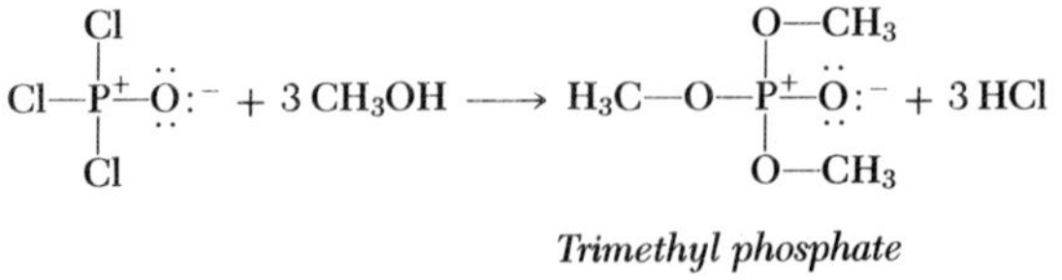

Trimethyl phosphate

Of equal interest to the biochemist are the condensation polymers of phosphoric acid. Heating phosphoric acid yields a mixture of anhydrides, including a monoanhydride known as pyrophosphoric acid ($H_4P_2O_7$), triphosphoric acid, and higher anhydrides, some of which are high-molecular-weight polymers (known collectively as metaphosphoric acid, HPO_3). Pyrophosphoric acid can be readily coverted into its methyl ester by the methods outlined previously.

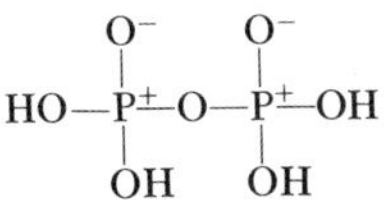

Pyrophosphoric acid 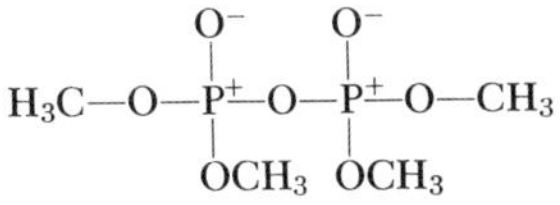   Tetramethyl pyrophosphate

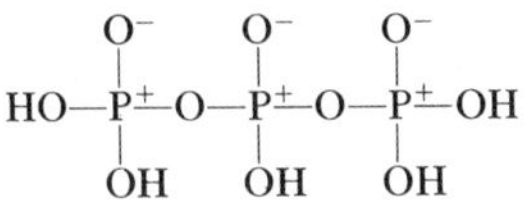

Triphosphoric acid

The methyl esters of the phosphoric acids resemble corresponding esters of sulfuric acid in reacting as alkylating agents.

Phosphoric acid forms derivatives of the carboxylic acids that are typical anhydrides; compounds of this type, such as acetyl phosphate, have also been isolated from natural sources. The reactions that these anhydrides undergo are similar to those of the corresponding acid chlorides (see p. 120 and the following reaction).

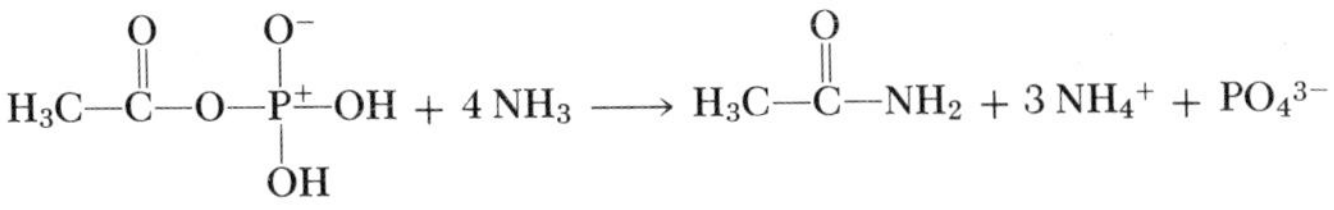

Acetyl phosphate

A number of derivatives of the carboxylic acids (RCOX) have been discussed (in this chapter and in Chapter 4) whose sole difference is the group X lost in the displacement reaction. It is instructive to see how the reactivity of these compounds with nucleophiles is related to the nature of group X (Figure 5.14). Notice that the more active the derivative is, the more weakly basic (and therefore the more stable) the leaving group (X) is. Almost the same series is involved in the reactivity of alkylating agents in nucleophilic displacement reactions, as we have see in the previous section (Figure 5.11).

The members on the right-hand side of Figure 5.14 have relatively high free energies and their reactions, for example with water, result in lower energy compounds. The net free-energy change is negative, and the equilibrium constant must be large (p. 51). Thus, the compounds with good leaving groups

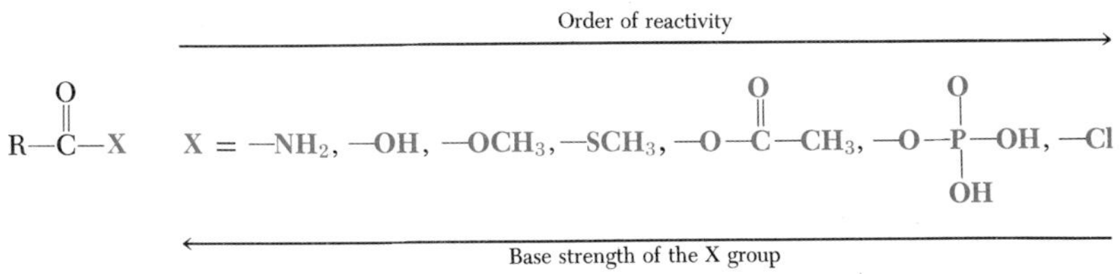

Figure 5.14 **The reactivity of various carboxylic acid derivatives.**

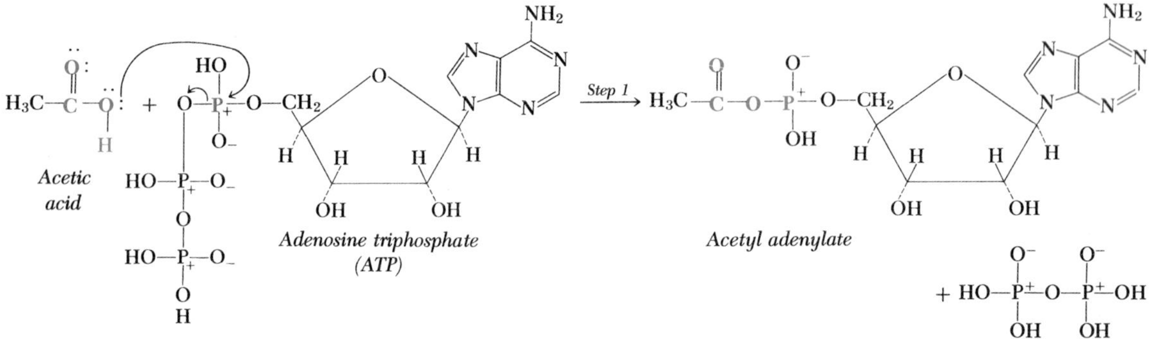

Figure 5.15 **Reactions of acid anhydrides and "activated" acids of biological interest (continued on the facing page).**

give high yields of products. In chemical terms, the lower the base strength of X^-, the slower it reacts in the reverse direction and thus the larger the equilibrium constant:

$$R-\overset{O}{\underset{\|}{C}}-X + CH_3O^-K^+ \rightleftharpoons R-\overset{O}{\underset{\|}{C}}-OCH_3 + K^+X^-$$

It should also be pointed out that there is an approximate relationship between the rate of a reaction and the position at equilibrium, so that, in general, the derivatives in the right-hand part of Figure 5.14 not only give a higher yield of product at equilibrium, but they also reach equilibrium at a faster rate than do those listed in the left-hand part. Biochemists often refer to compounds of the former category as "activated acids."

The two most important types of "activated acids" are (1) mixed anhydrides with phosphoric acid and (2) thiol esters. Examples of their reactions are given in Figure 5.15 [the nonionized acids are shown; however, in cellular fluids at pH 7, the OH groups attached to the phosphorus atoms would be fully ionized (see p. 121)]. The end product of the sequence, acetyl CoA, is an important intermediate for the incorporation of the acetyl group into citric acid (p. 122), fatty acids (p. 180), and other molecules of biological importance. Acetyl adenylate, ATP, and acetyl CoA are far more complex than are the simple molecules, such as acetyl chloride, that can undergo the same reactions in test tubes. However, in practice acetyl chloride is made and used under conditions that would be highly toxic to life, whereas the "activated acids" of Figure 5.15 work rapidly in water at pH 7 to give good yields of products. Under these conditions, a molecule such as acetyl chloride would merely react rapidly with the solvent to give acetic acid. The "molecular foliage" in the derivatives of Figure 5.15 is necessary for a good fit to the enzymes that catalyze the individual steps.

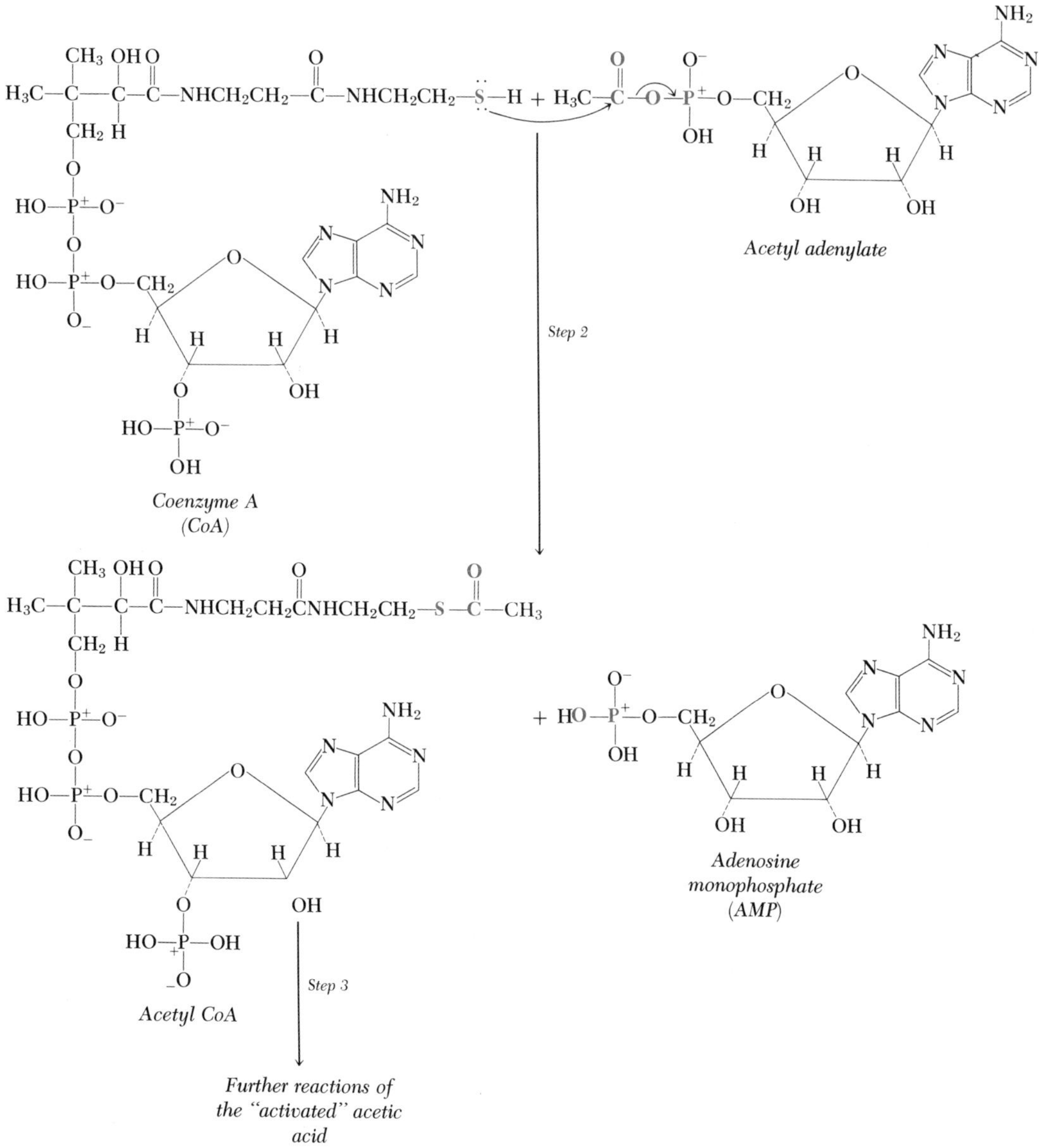

CH₃ OH O ... O

H₃C—C—C—C—NHCH₂CH₂—C—NHCH₂CH₂—S—H + H₃C—C—O—P—O—CH₂ ... NH₂

Acetyl adenylate

Step 2

Coenzyme A
(CoA)

H₃C—C—C—C—NHCH₂CH₂CNHCH₂CH₂—S—C—CH₃

+ HO—P—O—CH₂

Adenosine
monophosphate
(AMP)

Acetyl CoA

Step 3

Further reactions of
the "activated" acetic
acid

One of the unsolved problems of biology is how the enzymes specifically react with molecules of this type in water to undergo not simple hydrolysis, but to form new molecules containing new C—C bonds.

6 NATURAL PRODUCTS

NATURAL PRODUCTS ARE COMPOUNDS DERIVED FROM LIVING OR ONCE-LIVING organisms. An infinite variety of natural products has been isolated, but we can group most of these compounds into about a dozen classes. Of this dozen, three classes—proteins, carbohydrates, and fats—are especially important since they are the three main types of foodstuffs; principal emphasis in this chapter will be on these natural products.

PEPTIDES AND PROTEINS

Proteins are probably the most important of the chemical constituents of living organisms. They are the chief component of muscle fibers, skin, tendons, nerves, blood, and so on. In addition, enzymes, antibodies, and certain hormones are proteins. The proteins have been investigated intensively for many years and they may be defined, in a chemical sense, as high-molecular-weight polymers in which the building blocks are the amino acids.

The acid-catalyzed hydrolysis of proteins yields predominantly a mixture of the amino acids. Chromatographic methods are available today for the separation of the amino acids in mixtures of this type, and for the determination of the ratios in which the amino acids occur in the protein. It is interesting to note that each protein is composed of a different assortment of amino acids

Table 6.1 *The distribution of amino acids in certain proteins (Grams of amino acid/100 g protein)*[a]

PROTEIN	GLYCINE	ALANINE	VALINE	LEUCINE	ISOLEUCINE	METHIONINE	PHENYLALANINE	TRYPTOPHANE	LYSINE
Fibroin (silk)	44	30	4	1	1	—	3	—	1
Keratin (wool)	7	4	5	11		1	4	2	3
Albumin (hen)	3	7	7	9		5	8	1	6
Hemoglobin (horse)	6	7	9	15	7	1	8	2	9
Insulin (ox)	4	5	8	13	3	—	8	—	3

[a] Adapted from L. F. Fieser and M. Fieser, *Advanced Organic Chemistry* (New York: Reinhold Publishing Corp., 1961), p. 1025.

and that these are linked together in a unique sequence. The amino acid compositions of certain proteins are given in Table 6.1.

The fact that proteins are hydrolyzable to amino acids gives us no clue, of course, as to how the amino acids are bound together in the protein. Modern physical methods of analysis (principally infrared spectroscopy), on the other hand, have shown quite clearly that in a protein the nitrogen atom of one amino acid is bonded to the carbonyl group of another (forming an amide linkage) and that the carbonyl group of the first amino acid is bonded to the amino group of the third amino acid, and so on:

```
      H  H  O    H  H  O    H  H  O
      |  |  ||   |  |  ||   |  |  ||
···—N—C—C—[ N—C—C ]—N—C—C—···, etc.
         |        |        |
         R        R        R
```

The proteins are polymers of the amino acids, and the acid hydrolysis of the proteins, therefore, involves merely the hydrolysis of typical amide linkages. The polymers made up of a small number of amino acids (from two to about fifty) are commonly called *peptides*, whereas those formed of a larger number of units (up to 10,000 units or more!) are called *proteins*.

Because of their relatively small size, it is fairly easy to determine the structures of peptides and to synthesize them. The structure of a cyclic peptide containing eight amino acids is given in Figure 6.1. This compound, oxytocin, is highly active in regulating uterine contraction in childbirth. It was synthesized recently in the laboratory from the individual amino acids by essentially a stepwise procedure, in which the amino group of one amino acid was reacted with the activated-acid form of another amino acid (Chapter 5). The biosynthesis of proteins is similar in outline; the activated form is usually a mixed anhydride of the amino acid and adenosine monophosphate (AMP; see p. 157). The sequence of the different amino acids in the growing protein chain is determined ultimately by the DNA in the chromosomes of the cell; it is believed that one gene determines the sequence for one polypeptide. Further details

NATURAL PRODUCTS

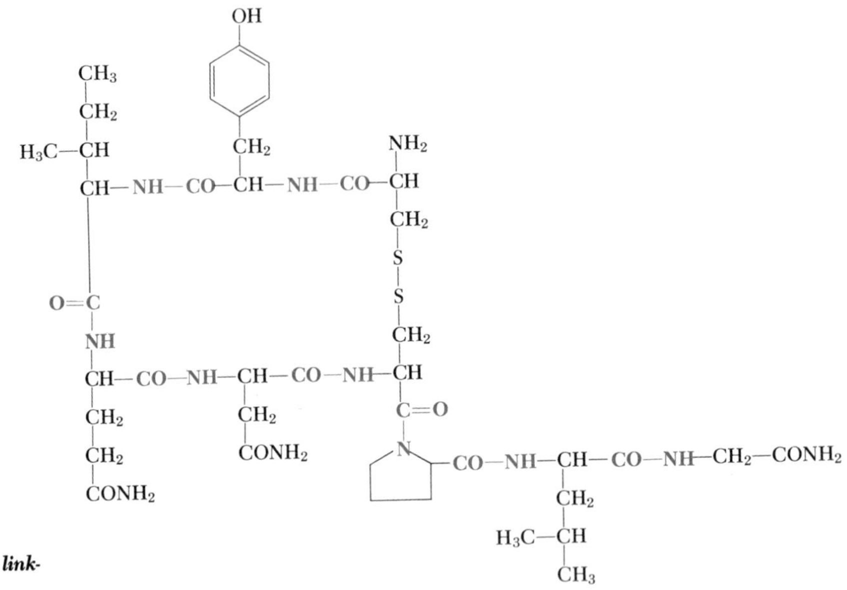

Figure 6.1 Oxytocin (the peptide linkages are given in color).

about protein structure and synthesis are given in other volumes of this series.

ENZYMES   One class of proteins is of special interest to chemists because of the ability of the members to increase the rate of reactions. These compounds, called enzymes, are catalysts that regulate the many chemical reactions occurring in a living organism; several hundred have been isolated, some in crystalline form, and each has a role to play in the catalysis of a particular reaction or set of reactions.

The molecular weights of the enzymes range from about 10,000 to several million. From the molecular weights of the amino-acid building blocks, it can be calculated that each enzyme molecule contains anywhere from 75 to tens of thousands of amino acids. Since each molecule may contain 20 kinds of amino acids, it is clear that 10,000 amino acids of 20 different kinds can be assembled in an extremely large number of ways. Enzyme molecules do not contain a random assortment of amino acids, however. They have a single specific sequence. Furthermore, all the molecules have exactly the same length of chain. These properties simplify somewhat the elaboration of the structure of the enzymes.

The enzyme papain, obtained from the papaya tree, catalyzes the hydrolysis of other proteins (it is used commercially to tenderize meats). Papain has a molecular weight of 21,000, and each molecule contains 185 amino acid residues distributed as follows: 23 glycine, 17 tyrosine, 17 glutamic acid, 17 aspartic acid, 15 valine, 13 alanine, 11 serine, 10 arginine, 10 leucine, 10 isoleucine, 9 lysine, 9 proline, 7 threonine, 6 cysteic acid, 5 tryptophane, 4 phenylalanine, and 2 histidine. These 185 amino acids are arranged in a single linear peptide chain and the specific sequence of amino acids is known for this enzyme. Most compounds with long chains of atoms have rubbery, noncrystalline properties, which are a reflection of the ability of long, coiled molecules to stretch (natural rubber and the synthetic varieties are examples). Papain, however, as well as many other enzymes, is a crystalline compound, and it is unlikely that a compound that really existed with a long stretched-out molecular chain would be crystalline. Other physical techniques have shown that the peptide chain is folded in a specific way—to give a more spherical molecule. Spherical molecules pack easily in a crystal lattice and compounds made up of such molecules are normally found in the crystalline state. The folding and shaping of peptide chains can occur by several different processes. Often the —SH bonds of two different cysteine molecules in the chain are oxidized, and an S—S bond is formed. This process leads to the formation of rings of atoms (see Figure 6.1 for an example). Hydrogen-bond formation between the amino group of one amide bond and the carbonyl group of another

represents another method for weak bonding within a single molecule of a protein. Such hydrogen bonding often forces the long peptide chain to adopt a helical shape. Still another type of intramolecular bonding results from the attraction of oppositely charged groups along the chain (for example, $-NH_3^+$ and $-CO_2^-$ groups). The interactions just discussed lead to the folding of the peptide chain in a very specific way. The geometry of the resulting molecule is intimately connected with the question of how the enzymes perform their catalysis of chemical reactions, in regard to both the specificity of the catalysis and the high rates of the reactions.

Most enzymes are highly specific in their action; the enzyme urease catalyzes only the hydrolysis of urea, for example, and the enzyme fumarase catalyzes the addition of water to fumaric acid (to yield malic acid), but it does not catalyze the addition of water to its geometrical isomer, maleic acid (the formulas of these acids are given in Figure 4.10). The mechanism of enzyme catalysis is not known with certainty at the present time, although considerable progress has been made recently, especially in regard to the complete three-dimensional structure as determined by X-ray crystallographic analysis. In general terms, it appears that the catalytic process involves (1) binding of the reactants in cavities in the folded enzyme by hydrogen bonding, charge attraction, and so on; (2) catalysis of bond-making and bond-breaking by hydrogen bonding and by charge interactions like those discussed as acid-catalyzed processes in chemical systems (Figures 3.20 and 4.5); (3) rapid ejection of the products. The cavity in which the critical catalysis occurs is called the active site of the enzyme; one of the goals of chemistry is the determination, in detail, of the structure of these active sites. With this information, it may be possible to synthesize small molecules that will be just as active catalytically as are the natural enzymes. The availability of such catalysts should make it much easier for the chemist to synthesize complex drugs, antibiotics, hormones, and so on, which might have a profound influence on medical practice.

The importance of enzymes to medicine is manifest when we consider hereditary diseases. A number of disorders of man have been traced to the inability of the body to produce one particular enzyme. Such diseases include familial goiter, hemophilia, albinism, and maple syrup urine disease. Once the causative factors have been established, the possibility exists that the disease can be checked or cured, either by administration of the enzyme or by correcting the conditions that led to the absence or shortage of the active form of the enzyme. At some time in the future, possibly even the genetic factors that cause these diseases may be corrected. More detailed accounts of the mode of action of the enzymes in various biological processes are given in the other volumes of this series.

One curious fact about proteins and amino acids remains to be discussed. It has been found that the amino acids isolated from the hydrolysis of proteins are the same as the *synthetic* amino acids with *one important exception:*

$$CH_3CH_2CO_2H \xrightarrow{Cl_2} CH_3CHClCO_2H \xrightarrow{NH_3} CH_3\overset{|}{C}HCO_2H$$

Propionic acid
$$NH_2$$

Synthetic alanine

The amino acids isolated from proteins rotate the plane of plane-polarized light, whereas the corresponding synthetic amino acids do not rotate this plane! This difference in behavior has its origins in a type of isomerism called optical isomerism. This topic will be outlined in the next section, where we shall see that there are two kinds of amino acids—those that rotate the plane of polarized light to the right, and those that rotate the plane to the left. An equimolar mixture of these forms, furthermore, does not rotate this plane. In living systems, the behavior of the two forms is quite different: one form may be an active metabolite, whereas the other form may be inactive or even toxic to the system.

The study of optical isomerism began with the observation that most compounds isolated from natural sources are able to rotate the plane of polarized light. Plane-polarized light may be obtained by passing ordinary light through certain minerals such as Iceland spar (a form of $CaCO_3$). Ordinary visible light is a form of electromagnetic radiation, which is usually regarded as a wave

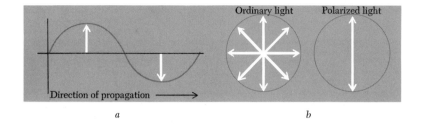

Ordinary light   Polarized light

Direction of propagation ⟶

a   *b*

Figure 6.2   (a) **A wave of light.** (b) **The magnetic vectors of beams of ordinary and plane-polarized light.**

phenomenon. The variation of the magnetic field associated with a single photon is given in Figure 6.2a. A beam of light contains a large number of photons, and if the beam is examined end-on, the oscillations of the magnetic field (represented by vectors) would be oriented in all possible directions, as in, Figure 6.2b (ordinary light). A properly prepared section of Iceland spar has the ability to transmit only light that has these vectors oriented in one particular direction (Figure 6.2b [polarized light]); light of this type is called plane-polarized light.

Materials such as Iceland spar may be used not only to form plane-polarized light, but also to determine any change in the inclination of this plane that may be brought about by amino acids and other natural products; the instrument used to measure these rotations is called a *polarimeter* (Figure 6.3). The polarizer and analyzer are sections of Iceland spar that transmit plane-polarized

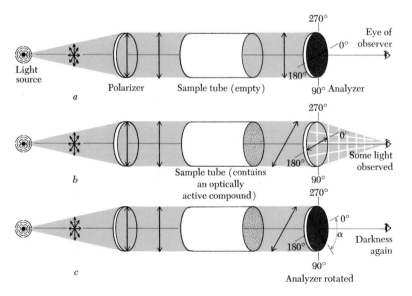

Figure 6.3   **Schematic diagram of a polarimeter.**

NATURAL PRODUCTS

light in one direction. If the sample tube in the polarimeter is empty, a maximum amount of light will reach the observer when the polarizer and analyzer are aligned and are passing light vibrating in the same plane. If the analyzer is turned, less and less light will reach the observer, until a minimum is reached, at which point the analyzer is at right angles to the polarizer and the field will be dark (Figure 6.3a). This is our starting point (0° on the scale). Now if an optically active compound such as an amino acid is introduced into the sample tube, the plane of the polarized light will be tilted by the compound (to the right in Figure 6.3b) and a certain amount of light will pass through the analyzer to the observer. The analyzer in now rotated a few degrees to the right (as in Figure 6.3c) until the field is dark again; at this point, the plane of the polarized light is again at right angles to the analyzer. The analyzer, then, has been turned a certain number of degrees (α), which is equal to the degree of tilting of the polarized light by the compound in the sample cell. Since α is dependent on the concentration of the compound in the sample tube and also on the length of the sample tube, optical activity is usually reported in terms of the specific rotation, $[\alpha]_D$, which is independent of these variables but is characteristic of a given compound: $[\alpha]_D = \alpha/lc$. Where α equals the observed reading, l equals the length of the cell in decimeters, c equals the concentration in grams per cubic centimeter, and the subscript D refers to the wavelength of the polarized light used.

If the compound rotates plane-polarized light to the right, and the analyzer must be moved to the right (or in a clockwise direction), the compound is said to be *dextrorotatory* (or $+$), and if it rotates plane-polarized light in the opposite direction, it is said to be *levorotatory* (or $-$). In either case, the compounds are called *optically active*. The specific rotations ($[\alpha]$) of optically active compounds range from values just above the error of the instrument (about 0.01°) to values of thousands of degrees. It should be emphasized at this point that, in general, most of the compounds that chemists use (such as ethanol, benzene, and so on) are optically inactive; that is, they do not rotate the plane of polarized light.

OPTICAL ACTIVITY   The ability of a compound to rotate the plane of plane-polarized light has been traced to the symmetry of the molecules of that compound. We can illustrate this relationship with the aid of molecular models; we recommend that the reader construct models to aid in visualizing the following material.

Suppose, for example, that we were to construct a ball-and-stick model of a carbon atom bearing two different types of substituents (as in CH_3Cl). We would find that we could superimpose the mirror image on this model (this merely requires a movement of the mirror image to the left until it coincides with the model as in Figure 6.4), and that there is only one model of this compound that can be constructed (Figure 6.4). The various models represented in Figure 6.4 are identical; for example, b may be converted into a by a rotation of 120° about the Y axis. Simple rotations of the other models will show that

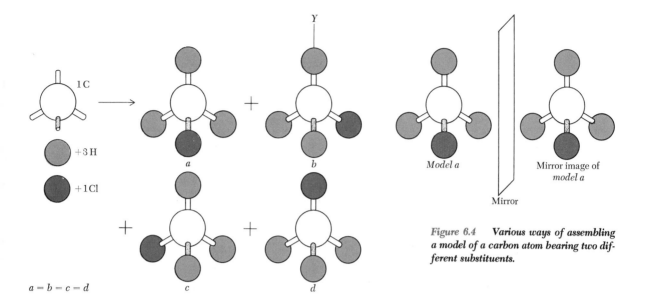

Figure 6.4 *Various ways of assembling a model of a carbon atom bearing two different substituents.*

$a = b = c = d$

forms c and d are also identical with form a. The situation is similar if three different substituents are attached; for example, a model of CH_2ClBr can be superimposed on its mirror image and only one model of the compound can be constructed. In this connection, it has been determined in the laboratory that the compounds CH_3Cl and CH_2ClBr are not optically active.

If the carbon atom bears four different substituents, however (as in CHClBrI), a model of the compound is not superimposable on its mirror image (Figure 6.5).

Figure 6.5 *Model, and its mirror image, of a carbon atom bearing four different substituents.*

Mirror

Model a

Mirror image of model a

C

I

Br

Cl

H

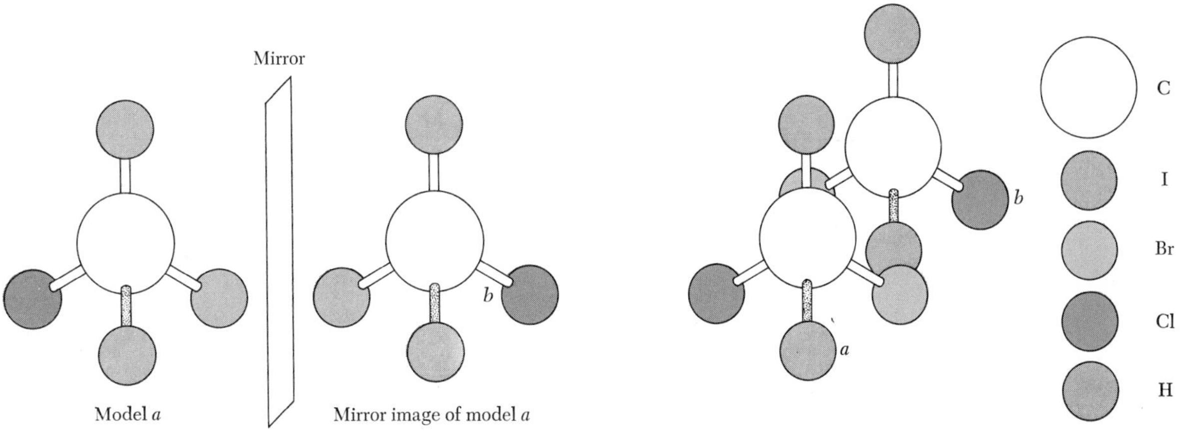

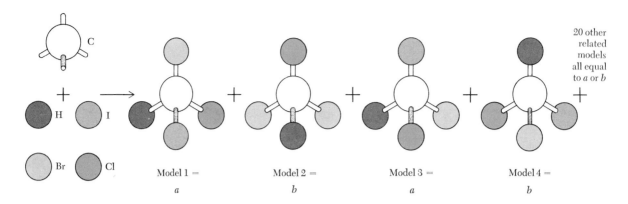

20 other related models all equal to *a* or *b*

Model 1 =
a

Model 2 =
b

Model 3 =
a

Model 4 =
b

Figure 6.6 Molecular models of **CHClBrI** *which show that only two forms of this compound can be constructed.*

Although we can superimpose the models so that the carbon, hydrogen, and iodine atoms coincide, as in *b*, the chlorine and bromine atoms occupy different regions in space; by no amount of rotation can the two models be superimposed. Furthermore, two different models of CHClBrI can be constructed (Figure 6.6), but only two. Furthermore, one of these two forms turns out to be identical to the model *a* in Figure 6.5, and the other is identical to the mirror image of model *a*. A simple but useful rule concerning these comparisons is that any model of a carbon atom bearing four different substituents is converted into its mirror image when the positions of any two substituents are exchanged.

Precisely the same behavior occurs at the molecular level. We find that there is only one kind of CH_3Cl (and of CH_2ClBr), but two kinds of CHClBrI. We can separate these two kinds (corresponding to model *a* and its mirror image in Figure 6.5) and study them independently. If we do, we find that they differ only by the way in which they rotate the plane of plane-polarized light. One form is dextrorotatory, the other is levorotatory. The *magnitude* of the rotation is the same, and therefore an equimolar mixture of the two forms has a specific rotation of zero. The two optically active forms are called *enantiomers*— which is a general term used to designate mirror-image forms that are not super-imposable—and the fifty-fifty mixture is called a *racemic mixture*. A carbon atom bearing four different substituents is called an *asymmetric* carbon atom and it can lead to a maximum of two optical isomers. Most compounds bearing an asymmetric carbon atom are optically active, although there are a few exceptions to this rule. The *maximum* number of optical isomers possible in com-pounds bearing more than one asymmetric carbon is given by the term 2^n, where n is the number of asymmetric carbon atoms in the molecule (however, in a few instances of cyclic or highly symmetrical compounds, fewer isomers than predicted by this rule may exist).

Optical isomerism is the third type of isomerism we have outlined; in Chap-ter 3, we covered geometrical isomerism and structural isomerism. These three types together make up the science of *stereoisomerism,* the study of the distribution of atoms in space.

Symmetry   It is often inconvenient to carry out a determination of whether a molecule is superimposable on its mirror image, therefore a shortcut has been devised for determining whether or not a compound is capable of optical activity. If a molecule has a plane of symmetry or a center of symmetry, it is not optically active, and it is called a *symmetric* molecule. If a molecule does not have these elements of symmetry, on the other hand, in the vast majority of cases it is optically active, and the molecules are called *asymmetric* compounds. A *plane of symmetry* is defined as a plane (conveniently visualized as a mirror) cleaving a molecule in such a way that one side of the molecule is a mirror image of the other. A *center of symmetry* is defined as a point (at the center of a molecule) which is so situated that any straight line through it passes through the same environment in both directions extending from that point.

We can illustrate these elements of symmetry with the aid of a few common objects (Figure 6.7). The average coffee pot (*a*) has a plane of symmetry that bisects the handle, pot, and spout. A specially constructed coffee pot with the spout at 90° from the handle (*b*) would have no symmetry elements, and it could be termed an asymmetric pot. A cube (*c*) is a highly symmetric object with a center of symmetry and several planes of symmetry (nine, to be exact). A cube bearing four different pairs of balls at the corners, arranged as in Figure 6.7*d*, on the other hand, has a center of symmetry as its only symmetry element.

Living things also possess certain symmetry elements; actually most animals and most plants (other than those with alternate leaves) possess planes of symmetry. To return to the molecular level, examples of optically active compounds are given in Figure 6.8, and examples of inactive compounds are given in Figure 6.9, to illustrate the symmetry rules.

In calculating the number of isomers a compound has, we assume that there is free rotation about single bonds in noncyclic compounds. Usually that conformation is chosen which has the maximum amount of symmetry; that is, the compound CH_2ClCH_2Cl is not optically active although a casual examination

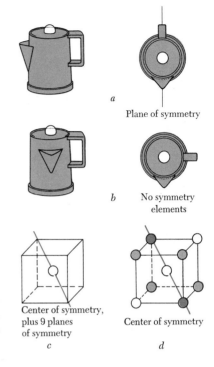

a

Plane of symmetry

b   No symmetry elements

Center of symmetry, plus 9 planes of symmetry

c

Center of symmetry

d

Figure 6.7   Symmetry elements of common objects.

Alanine

H_3C—$\overset{\overset{\displaystyle H}{|}}{\underset{\underset{\displaystyle NH_3^+}{|}}{C}}$—$CO_2^-$

Alanine
2 optical isomers (+ and −)
($2^n = 2$)
$[\alpha]_D = +8.5°$ and $−8.5°$

H_3C—$\overset{\overset{\displaystyle H}{|}}{\underset{\underset{\displaystyle OH}{|}}{C}}$—$\overset{\overset{\displaystyle H}{|}}{\underset{\underset{\displaystyle NH_2}{|}}{C}}$—$CH_3$

3-Amino-2-butanol
4 optical isomers
($2^n = 4$)
2 + isomers and
2 − isomers

Figure 6.8   Optically active compounds.

NATURAL PRODUCTS

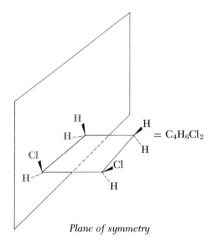

$$CH_2ClBr = $$

Br, H, C, H, Cl (structure shown)

Plane of symmetry is the plane of paper, which bisects the Br and Cl atoms and the angle HCH

Cl, H, Br, H, Br, H, Cl structure $= C_4H_4Br_2Cl_2$

Center of symmetry

$= C_4H_6Cl_2$

Plane of symmetry

Figure 6.9 **Optically inactive compounds.**

might suggest that it is.

Cl, H, H, C—C, Cl, H $\rightleftharpoons$ Cl, H, H, C—C, H, Cl structure

Conformation a   Conformation b

1,2-Dichloroethane

Conformation *b* has a plane of symmetry in the plane of the paper.

Resolution   If enantiomers have exactly the same chemical properties, the question arises, how can they be separated? Several methods have been developed, but the most common procedure depends on the three-dimensional structure of asymmetric compounds. If a pure enantiomer of an amine is obtained from natural sources—for example, (−) 2-butylamine (CH_3CHNH_2-CH_2CH_3)—it can be reacted with a racemic mixture of (+) and (−) alanine; the result is a mixture of salts, necessarily formed in equal molar quantities. It can be seen from the three-dimensional representation in Figure 6.10 that the salts are made up of different partners. The forms shown will have different rotations, and the (+ −) form will have a higher rotation than will the (− −) form. Molecules that contain two or more asymmetric centers and are not enantiomeric (that is, are not mirror images) are called *diastereomers*. The enantiomer of Figure 6.10*a* would be the (− +) salt prepared from the (−) acid and the (+) amine, and the enantiomer of Figure 6.10*b* would be the (+ +) salt prepared in turn from the + and + components.

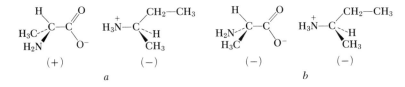

Figure 6.10   *The two diastereomeric salts of (—)2-butylamine and (+) and (—) alanine.*

Diastereomers (for example, Figure 6.10*a* and *b*) have different physical properties. Usually one form is less soluble than the other, and on fractional crystallization one diastereomer can usually be obtained pure. If, for example, the (+ —) 2-butylammonium alanate were less soluble than the (— —) form, it would crystallize out of the reaction mixture first; treatment of this salt with concentrated base would then liberate the free amine:

$$
\underset{(+)}{H_3C-\overset{\overset{\displaystyle H}{|}}{\underset{\underset{\displaystyle NH_2}{|}}{C}}-CO_2^-} \quad \underset{(-)}{\overset{+}{H_3N}-\overset{\overset{\displaystyle C_2H_5}{|}}{\underset{\underset{\displaystyle H}{|}}{C}}-CH_3} \xrightarrow{K^+OH^-}
$$

$$
\underset{(+)}{H_3C-\overset{\overset{\displaystyle H}{|}}{\underset{\underset{\displaystyle NH_2}{|}}{C}}-CO_2^-K^+} + \underset{(-)}{H_2N-\overset{\overset{\displaystyle C_2H_5}{|}}{\underset{\underset{\displaystyle H}{|}}{C}}-CH_3} + H_2O
$$

The amine is very volatile, and it could be removed by distillation. The residual potassium salt would then be treated with a strong acid to yield (+) alanine, a pure enantiomer of alanine. The (—) form can usually be obtained in a similar way from the more soluble diastereomer. In effect, by this procedure the optically inactive mixture has been separated, and the two pure enantiomers isolated. This separation of a racemic mixture into its constituent enantiomers is called *resolution*.

A second method of resolution depends on the selectivity of enzymes. Most enzymes discriminate between optical enantiomers. Thus, if racemic alanine is acted on by an enzyme such as transaminase, the (+) alanine would be converted into pyruvic acid by the reverse of steps 1 to 3 on pp. 146–147; the (—) alanine would be untouched and could be isolated from the reaction mixture at the end of the experiment. By the use of these two methods of resolution, a very large number of optically active compounds have been isolated and characterized.

Asymmetric induction   Materials prepared in the laboratory from optically inactive compounds are always racemic mixtures. For example, a racemic cyanide is obtained from the addition of HCN to methyl ethyl ketone:

$$HCN + H_2O \rightleftharpoons H_3O^+ + CN^-$$

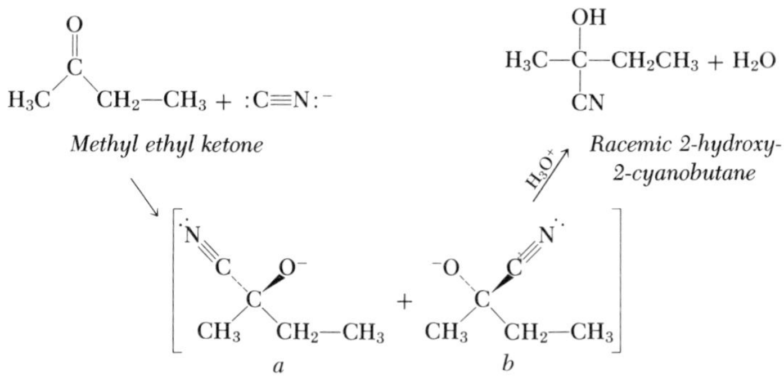

Methyl ethyl ketone

Racemic 2-hydroxy-2-cyanobutane

There is an equal probability that the cyanide ion will attack from the "back" or from the "front" (defined in terms of the plane of the paper) of the essentially planar methyl ethyl ketone molecule. This means that ions a and b are formed in equal quantities: ions a and b are enantiomers, and the protonation of these species then yields racemic 2-hydroxy-2-cyanobutane.

For similar reasons, a symmetric molecule will react with enantiomers at exactly the same rate. For example, when racemic alanine is treated with sodium hydroxide, both the $(+)$ and the $(-)$ enantiomers will react with hydroxide ion at the same rate to form the sodium salt of alanine.

A completely different picture emerges if we examine reactions of molecules already bearing one asymmetric center, such as the reaction of cyanide ion with the pure $(+)$ enantiomer of methyl-2-butyl ketone (the asymmetric carbon atom is starred):

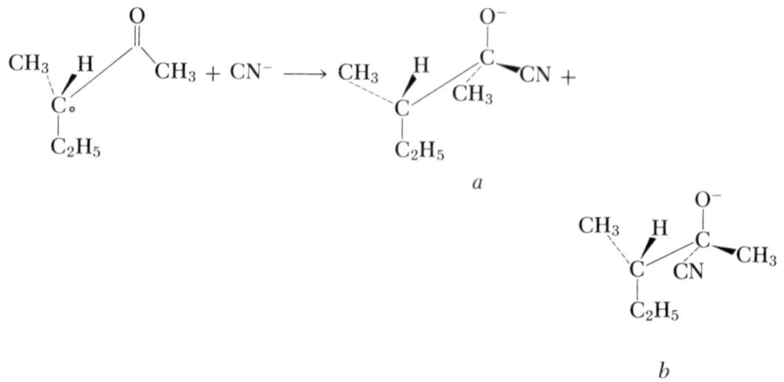

In this case, the approach of cyanide ion from the same side as the proton on the starred atom leads to a whereas approach from the side bearing the methyl group leads to *b*. Since a methyl group is much larger than a proton it will interfere with the approach of the cyanide ion, and the addition will favor *a*. No matter how we rotate the starred atom, there will be a difference between attack on the "top" of the carbonyl group (leading to *a*), and attack on the "back" (leading to *b*), and one form will be favored over the other. In summary, when a new asymmetric center is generated in a molecule already containing an asymmetric carbon, the new center will contain more of one isomer than of another and partial resolution will be the overall result. This process is called *asymmetric induction*. Often, only a single isomer is formed; in the example cited above, isomer *a* is essentially the only one formed. Enzymes contain many asymmetric centers and when the enzymes react with planar or nonasymmetric molecules, the same phenomenon occurs; that is, the ultimate product of the catalyzed reaction is only one of several possible isomers. For example, the following sequence is catalyzed by the enzyme aconitase as a part of the citric acid cycle:

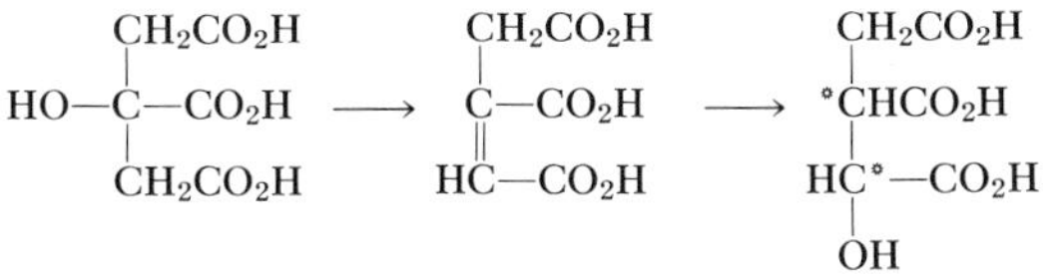

$[\alpha]_D = +20.9°$ $[\alpha]_D = -20.9°$

a   *b*

Figure 6.11   **The enantiomers of gylceraldehyde.**

Citric acid   *cis*-Aconitic acid   Isocitric acid

Citric acid has a plane of symmetry (bisecting the $HO—C—CO_2H$ atoms, and with a CH_2CO_2H group on each side), and in simple chemical reactions it would give only racemic products. The enzyme aconitase, however, first binds or attaches itself to the citric acid and then performs the chemical transformation to give optically pure isocitric acid (that is, only one of the four possible isomers); the asymmetric carbon atoms are starred.

Configuration   If a compound bearing one asymmetric carbon atom, such as glyceraldehyde (α,β-dihydroxypropanal, Figure 6.11) can exist in either the (+) or the (−) form, does form *a* rotate the plane of polarized light to the right, and form *b* rotate the plane to the left? Or is the situation reversed? The answer is that form *a* is the enantiomer of glyceraldehyde that rotates the plane of polarized light to the right. This assignment was made arbitrarily at first by the German chemist Emil Fisher in 1891. Fisher had, of course, a fifty-fifty chance of being right, but his choice was proved to be the correct one in 1952 by Dutch scientists using a special type of X-ray analysis.

Optically active compounds can give a wide variety of optically active derivatives; an example is given in Figure 6.12. In the formation of the benzoate ester shown, no bond to the asymmetric carbon has been broken, and the ester has the same distribution of H, CHO, O^-, and CH_2O^- in space about the asymmetric carbon atom as does (+) glyceraldehyde. This derivative is said

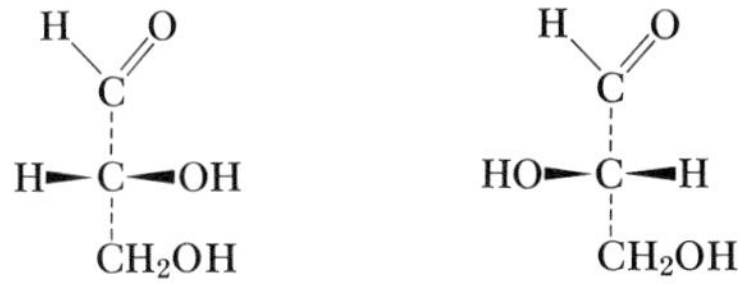

NATURAL PRODUCTS

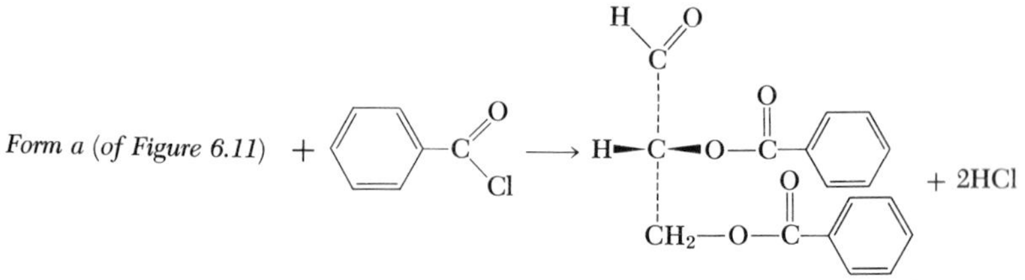

Form a (of Figure 6.11) + ⟶ + 2HCl

The dibenzoate of (+) glyceraldehyde
$[\alpha]_D = -35.6°$

Figure 6.12   *A levorotatory derivative of* D(+)*glyceraldehyde.*

The dibenzoate of (−) glyceraldehyde
$[\alpha]_D = +36.5°$

Figure 6.13   *A derivative of* L(−)*glyceraldehyde.*

to have the same *configuration* as (+) glyceraldehyde, where configuration is defined as the arrangement of atoms that characterize a stereoisomer.

All derivatives of (+) glyceraldehyde have the same configuration about the asymmetric carbon; they are said to belong to one family, which by convention is called the D family. Conversely, all derivatives of (−) glyceraldehyde belong to the L family. The symbols D and L are taken from the Latin *dextro* ("right") and *levo* ("left") referring to the signs of rotation, (+) and (−), of the two forms of glyceraldehyde. An example of a derivative of L(−)glyceraldehyde is given in Figure 6.13. Note that members of the same family often have rotations of opposite sign, or to state it differently, the sign of rotation does not identify what family a compound belongs to or what the distribution in space of the groups attached to the asymmetric carbon is. In these compounds (Figures 6.12 and 6.13), the configuration of the compound (that is, the distribution of groups in space) is given by the D or L label, whereas the (+) or (−) sign merely gives the direction of rotation of plane polarized light produced by the compound.

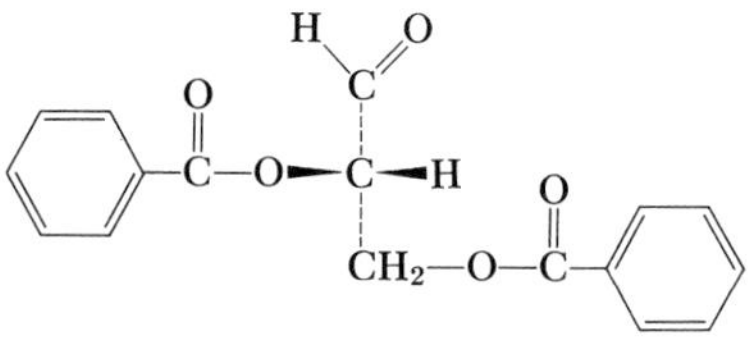

The configuration of an asymmetric carbon is usually determined experimentally by one of three methods. Method 1 (rarely used) is the absolute X-ray method mentioned previously. Method 2 is based on the principle that if a derivative of an asymmetric compound is prepared without breaking a bond to the asymmetric carbon atom in the reference compound, then the derivative belongs to the same family as does the reference compound. This was true of D-glyceraldehyde and the benzoate illustrated in Figure 6.12. Further examples are given in Figure 6.14; all of the compounds illustrated have the D configuration.

In method 3, bonds to the asymmetric carbon atoms are made—and broken—but it is known from other lines of research whether the new group comes in from the same side of the asymmetric carbon atom as the leaving group, or whether it comes in from the opposite side. For example, it has been established that nucleophilic displacement reactions proceed with *inversion* of configuration—which means that the reaction of a D molecule yields an L product, or vice versa, of course (Figure 6.15). The species Z is the halfway

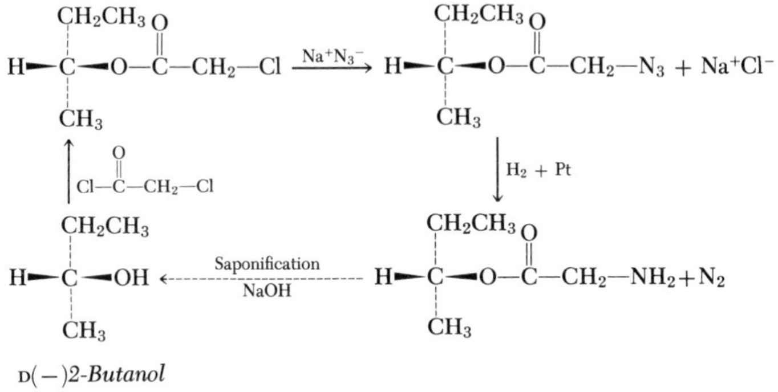

D(−)2-Butanol

Figure 6.14 **Derivatives of D(−)2-butanol.**

Figure 6.15 **The inversion of configuration attending the nucleophilic displacement of benzenesulfonate ion by bromide ion. (The dot-dash lines in compound Z represent partial bonds.)**

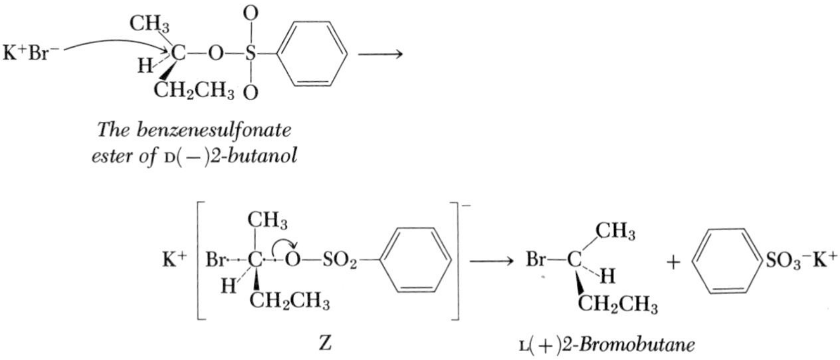

The benzenesulfonate ester of D(−)2-butanol

Z

L(+)2-Bromobutane

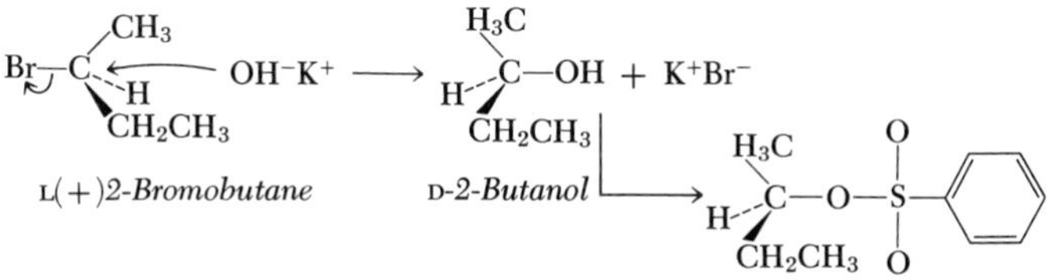

L(+)2-Bromobutane   D-2-Butanol

Figure 6.16 **Inversion of configuration in the displacement of bromide ion by hydroxide ion.**

Figure 6.17 **The formation of racemic products in a carbonium ion reaction: R =** ⬡

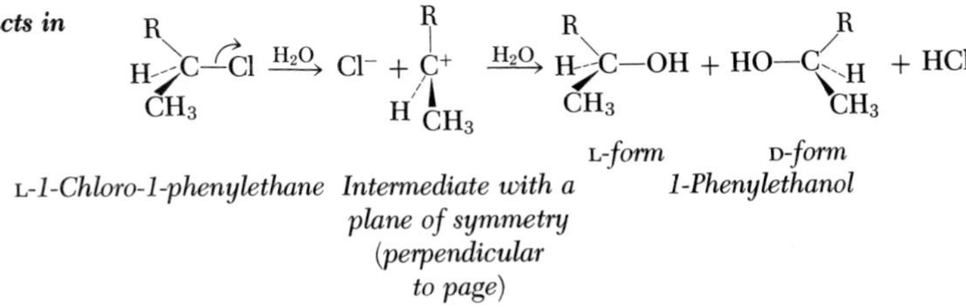

L-1-Chloro-1-phenylethane   Intermediate with a plane of symmetry (perpendicular to page)

L-form   D-form
1-Phenylethanol

173

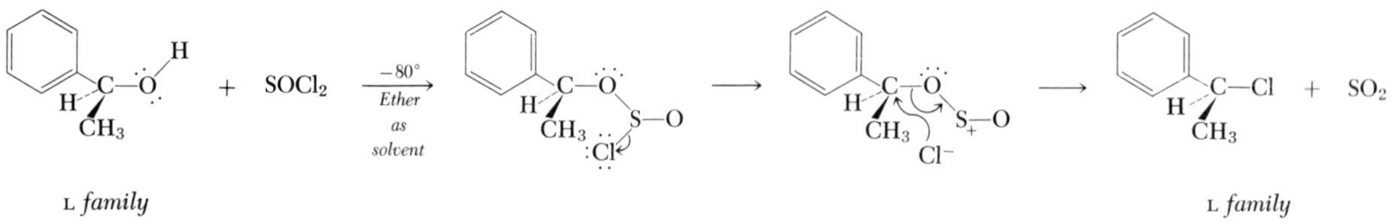

Figure 6.18   A reaction that proceeds
with retention of configuration.

point in the reaction, and the process shown in the figure for the inversion of the H, CH_3, and CH_3CH_2 groups resembles the movements of the ribs of an umbrella when they are inverted in a high wind.

The nucleophilic displacement of bromide ion from our compound by hydroxide ion will now give us the D alcohol (Figure 6.16) and this can be esterified to give us the derivative we started with in Figure 6.15. These transformations illustrate the rule that two inversions of configuration (once by the bromide ion and once by the hydroxide ion) have the same effect as one retention of configuration (that is, no net effect on the configuration).

Carbonium ion reactions, in contrast, normally lead to a loss of optical activity; that is, racemic or largely racemic products are usually obtained from reactions of this type (Figure 6.17). We can attribute the formation of these racemic products to the fact that carbonium ions are planar (because of the sp^2 hybridization of the electron-deficient carbon). A carbonium ion intermediate therefore possesses a plane of symmetry (in Figure 6.17, this plane is perpendicular to the plane of the paper), and any product formed from the carbonium ion must be optically inactive.

The study of optically active compounds is very important since it allows us to determine whether a reagent molecule has entered on the front of the molecule or on the back. Although most of the reactions that are known proceed with inversion of configuration (accompanied occasionally by more or less racemization) a few are known that proceed with retention of configuration (Figure 6.18).

CARBOHYDRATES

Carbohydrates are compounds of carbon, hydrogen, and oxygen derived more or less directly from carbon dioxide and water in photosynthesis. Sugar, starch, and cellulose are examples of carbohydrates that illustrate the importance of this class of compounds to life. Cellulose is the principal constituent of wood (and paper), starch is the principal constituent of grains and other seeds, and the sugars, in addition to their widespread use as a foodstuff, are important functioning parts of all living organisms. The term "carbohydrate" stems from the fact that many of the sugars have the empirical formula $C_nH_{2n}O_n$, which corresponds technically to a "hydrate of carbon," $C_n(H_2O)_n$.

CHEMICAL BACKGROUND FOR THE
BIOLOGICAL SCIENCES

Structurally, the sugars are hydroxylated aldehydes and ketones; if the compounds are single units they are called *monosaccharides*, whereas if more than one unit is present in the molecule, they are called *polysaccharides* (more specifically, *disaccharides* if two units are present, *trisaccharides* if three units are present, and so on). The saccharides are further broken down into *aldoses* and *ketoses*, depending on whether an aldehyde group or a keto group is present in the molecule. Also, the monosaccharides are subdivided to indicate the number of carbon atoms in the molecule. A *tetrose* contains four carbon atoms, a *pentose* has five, a *hexose* six, and so on.

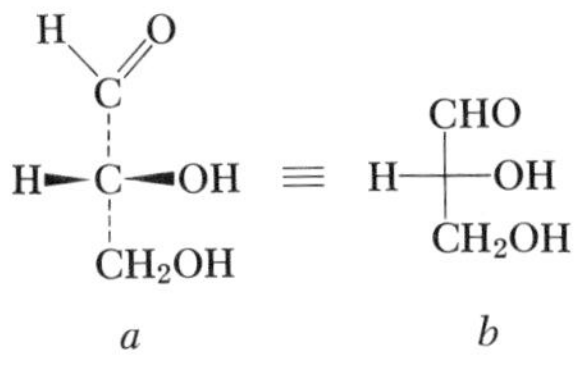

Figure 6.19 **A projection formula and a schematic diagram for D-glyceraldehyde.**

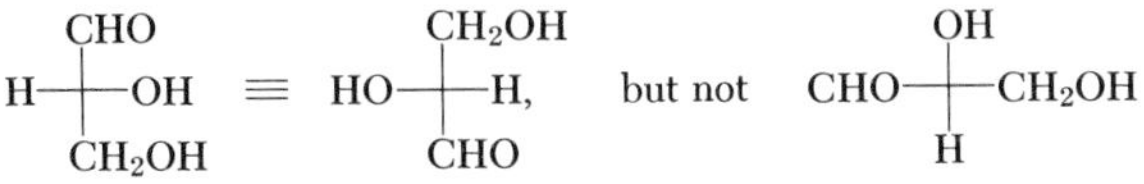

Figure 6.20 **Schematic diagrams of D(+)glyceraldehyde.**

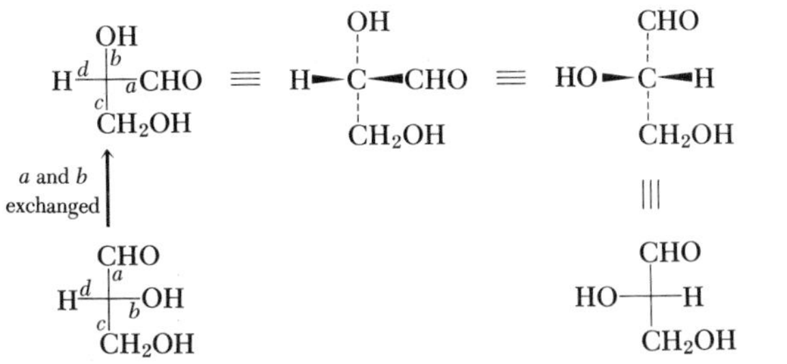

Figure 6.21 **An illustration of the fact that exchange of any two groups on an asymmetric carbon atom gives the enantiomeric configuration.**

D(+)Glyceraldehyde   L(−)Glyceraldehyde

MONOSACCHARIDES   In the previous sections, we discussed the stereochemistry of the aldotriose glyceraldehyde. This compound (Figure 6.11), which has only two stereoisomers, (+) and (−), plays a central role in assigning the configuration of the other saccharides. It is inconvenient to draw projection formulas (see Figure 6.19a) for complex compounds. Consequently, molecules of this type are often represented by schematic diagrams such as that shown in Figure 6.19b. In these diagrams, the vertical line represents bonds extending away from the reader (behind the plane of the paper) whereas the horizontal lines represent bonds coming out of the paper toward the reader. In working with these diagrams, certain rules must be followed: (1) the diagrams may be rotated in the plane of the paper by 180° but *not* by 90° (Figure 6.20), and (2) the exchange of any two groups gives the enantiomeric configuration for that particular asymmetric carbon atom (Figure 6.21). These rules should be verified through the use of ball-and-stick models. The schematic diagrams are useful in that they clearly illustrate mirror-image relationships (Figure 6.22).

Figure 6.22 **D- and L-glyceraldehyde. Each is the mirror image of the other.**

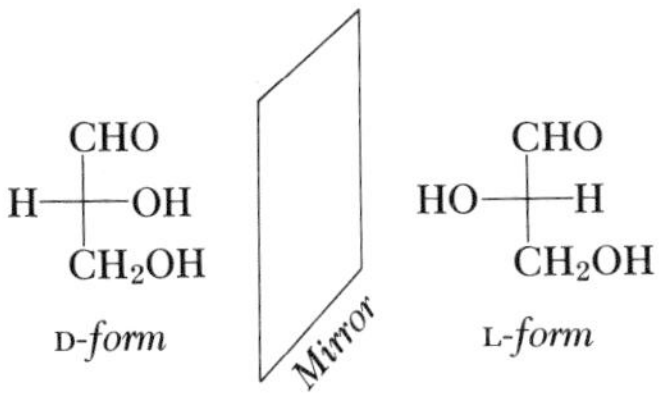

D-form   L-form

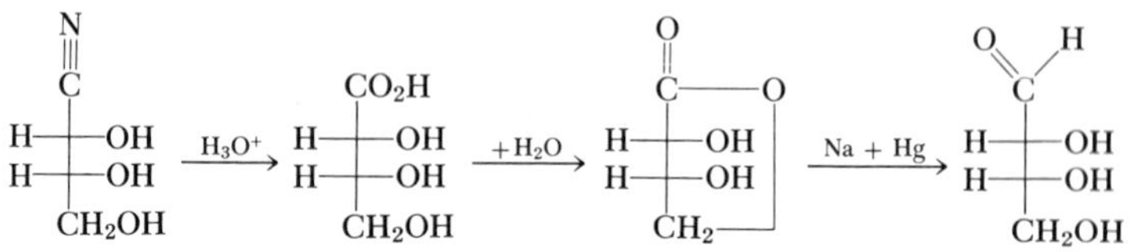

$$
\begin{array}{cccccccc}
& & & & \text{CO}_2\text{H} & & & \\
\text{N} & & & & | & & \text{O} & \text{O}\quad\text{H} \\
\text{|||} & & & & & & \text{||} & \diagdown\diagup \\
\text{C} & & & & & & \text{C}\!-\!\text{O} & \text{C} \\
\text{H}\!-\!\!-\!\text{OH} & \xrightarrow{\text{H}_3\text{O}^+} & \text{H}\!-\!\!-\!\text{OH} & \xrightarrow{+\text{H}_2\text{O}} & \text{H}\!-\!\!-\!\text{OH} & \xrightarrow{\text{Na}+\text{Hg}} & \text{H}\!-\!\!-\!\text{OH} \\
\text{H}\!-\!\!-\!\text{OH} & & \text{H}\!-\!\!-\!\text{OH} & & \text{H}\!-\!\!-\!\text{OH} & & \text{H}\!-\!\!-\!\text{OH} \\
\text{CH}_2\text{OH} & & \text{CH}_2\text{OH} & & \text{CH}_2\!-\!\!\! & & \text{CH}_2\text{OH} \\
\end{array}
$$

Figure 6.23 **The conversion of a cyanohydrin into an aldose.**

D-Glyceraldehyde reacts readily with HCN to give a mixture of two cyanohydrins; one new asymmetric carbon atom is generated in this reaction, as shown here:

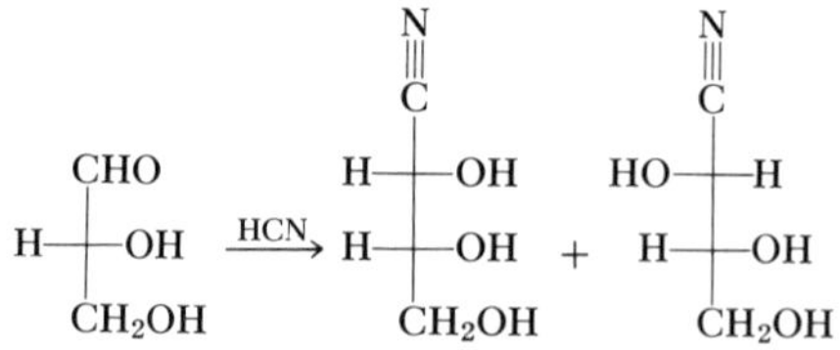

$$
\begin{array}{ccccc}
& & \text{N} & & \text{N} \\
& & \text{|||} & & \text{|||} \\
& & \text{C} & & \text{C} \\
\text{CHO} & & \text{H}\!-\!\!-\!\text{OH} & & \text{HO}\!-\!\!-\!\text{H} \\
\text{H}\!-\!\!-\!\text{OH} & \xrightarrow{\text{HCN}} & \text{H}\!-\!\!-\!\text{OH} & + & \text{H}\!-\!\!-\!\text{OH} \\
\text{CH}_2\text{OH} & & \text{CH}_2\text{OH} & & \text{CH}_2\text{OH} \\
\end{array}
$$

The ratio of isomers is not fifty-fifty because the glyceraldehyde molecule already contains an asymmetric carbon atom and the asymmetric forces set up with the incoming cyanide ion favor one form slightly over the other. In any event, each cyanohydrin can be converted into the corresponding tetrose (Figure 6.23).

By methods of this type, the four aldotetroses have been synthesized. Their configurations and names are given in Figure 6.24. The number of isomers could have been predicted from the expression (2^n) discussed in the previous section of this chapter, where n equals the number of asymmetric carbon atoms in the molecule.

It should be noted that, by definition, the family relationships of the higher saccharides are assigned on the basis of the configuration of the bottom-most asymmetric carbon atom in the schematic diagrams of the compounds (oriented so that the aldehyde group is at the top). That is, the tetroses with an OH group on carbon atom 3 on the right side of the formula are members of the D family. It can be seen from Figure 6.22 that the two tetroses synthesized from D-glyceraldehyde must be D- tetroses, whereas the two synthesized from L-glyceraldehyde must be L-tetroses.

By an extension of the cyanohydrin synthesis (Figure 6.23), the eight aldopentoses (Figure 6.25) have been synthesized from the tetroses. By similar procedures, furthermore, the 16 aldohexoses have been synthesized from aldopentoses. The three most common aldohexoses are given in Figure 6.26, along with the most common ketohexose, fructose.

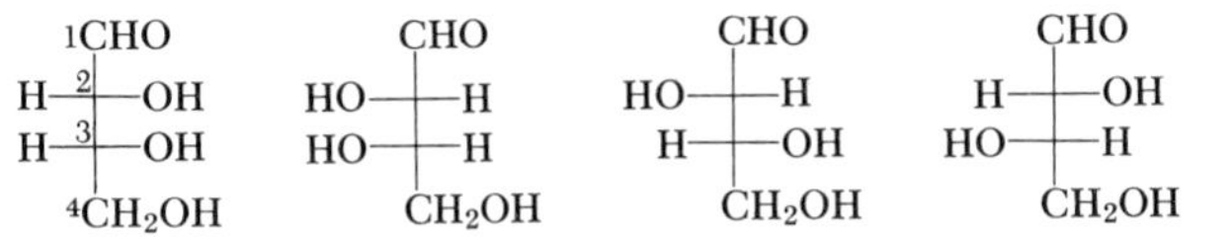

The aldotetroses.

D-Erythrose:
CHO
H—OH (2)
H—OH (3)
CH₂OH (4)

L-Erythrose:
CHO
HO—H
HO—H
CH₂OH

D-Threose:
CHO
HO—H
H—OH
CH₂OH

L-Threose:
CHO
H—OH
HO—H
CH₂OH

Figure 6.24 **The aldotetroses.**

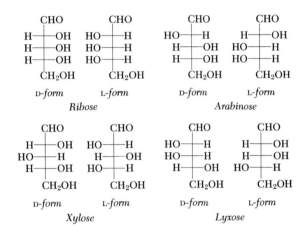

Ribose
D-form:
CHO
H—OH
H—OH
H—OH
CH₂OH

L-form:
CHO
HO—H
HO—H
HO—H
CH₂OH

Arabinose
D-form:
CHO
HO—H
H—OH
H—OH
CH₂OH

L-form:
CHO
H—OH
HO—H
HO—H
CH₂OH

Xylose
D-form:
CHO
H—OH
HO—H
H—OH
CH₂OH

L-form:
CHO
HO—H
H—OH
HO—H
CH₂OH

Lyxose
D-form:
CHO
HO—H
HO—H
H—OH
CH₂OH

L-form:
CHO
H—OH
H—OH
HO—H
CH₂OH

Figure 6.25 **The aldopentoses.**

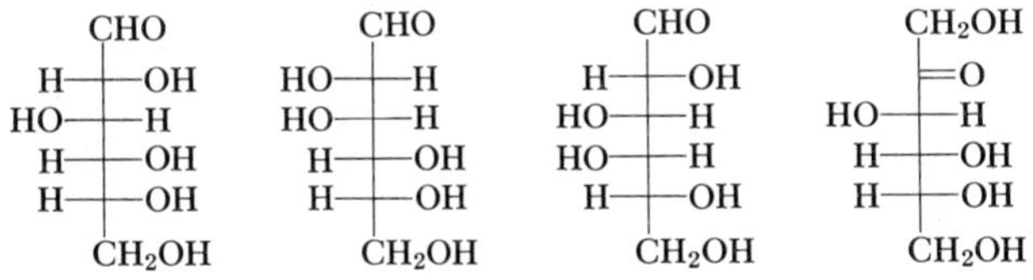

Three common aldohexoses and one ketohexose.

D-Glucose:
CHO
H—OH
HO—H
H—OH
H—OH
CH₂OH

D-Mannose:
CHO
HO—H
HO—H
H—OH
H—OH
CH₂OH

D-Galactose:
CHO
H—OH
HO—H
HO—H
H—OH
CH₂OH

D-Fructose:
CH₂OH
=O
HO—H
H—OH
H—OH
CH₂OH

Figure 6.26 **Three common aldohexoses and one keto-hexose.**

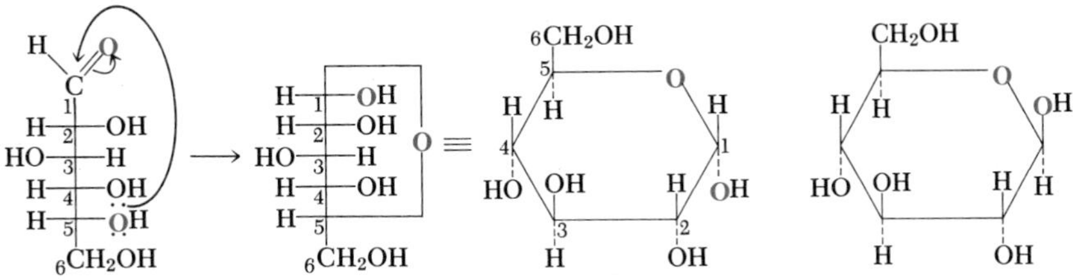

D-Glucose (open-chain form):
H—C=O (1)
H—OH (2)
HO—H (3)
H—OH (4)
H—OH (5)
CH₂OH (6)

α-form and β-form cyclic structures

Figure 6.27 **α-D-glucose and β-D-glucose.**

177

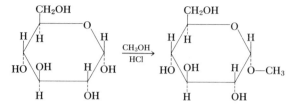

Figure 6.28   *α-Methyl glucosides.*

GLYCOSIDES   We have represented the pentoses and hexoses as linear molecules, but actually they exist largely in a cyclic hemiacetal form. Glucose, for example, has the structure shown in Figure 6.27, in which the OH group on carbon atom 5 has added to the carbonyl group. A new asymmetric center has been formed in this process, and the OH group on carbon atom 1 is either *cis* or *trans* to the OH group on carbon atom 2.

Both forms of glucose have been isolated; the former is called α-D-glucose and the latter β-D-glucose (Figure 6.27). In Chapter 2, mention was made of glucose-1-phosphate and glucose-6-phosphate. These compounds have structures based on α-D-glucose, in which the OH groups on carbon atoms 1 and 6, respectively, have been esterified with phosphoric acid ($-OH \rightarrow -O-PO_3H_2$).

As typical hemiacetals, the cyclic forms of the saccharides react with alcohols to form acetals (Figure 6.28). The acetals of sugars in general are called *glycosides*, although specifically the acetals of glucose are called *glucosides*. Glucose is widely distributed in nature, where it is found in the free state in honey and in ripe fruits, and in the bound state through a glucoside linkage to a wide variety of compounds.

The pentoses also exist largely in a cyclic hemiacetal form, and acetal formation can also occur. As in simple acetals, the bond to carbon atom 1 is easily broken to yield a carbonium ion; the ion can then react with various oxygen and nitrogen nucleophiles to give acetal-type compounds. The ATP and CoA molecules shown in Figure 5.15 contain ribose bonded to nitrogen in this way. Compounds of the ATP and CoA type involving a sugar molecule

Figure 6.29   **Sucrose.**

H—	—O—	
H—	—OH	
HO—	—H	O
H—	—OH	
H—		

CH₂OH

HO—	—H
	H—	—OH
H—	

CH₂OH

Glucose part   *Fructose part*

Sucrose

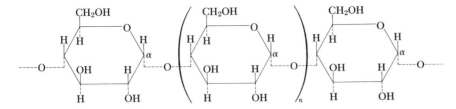

Figure 6.30   **A representative portion of a chain in the giant molecule of starch.**

esterified with phosphoric acid and also bonded to a heterocyclic compound are called *nucleotides*. The structural features of the nucleotides appear to be relatively important; in addition to ATP and coenzyme A, the compound DPNH (p. 100) is a nucleotide, and polymers of the nucleotides (RNA and DNA) are the transmitters of genetic information.

POLYSACCHARIDES   If the acetal of a monosaccharide is formed with the hydroxyl group of a second monosaccharide, a disaccharide results. The common "sugar" of commerce is the disaccharide called sucrose in which the hydroxyl group of a fructose molecule is used in forming the acetal of glucose (Figure 6.29). Trisaccharides, tetrasaccharides, and the higher saccharides, in which all combinations of the monosaccharides may be found, also have structures of this type.

The complete hydrolysis of starch or cellulose yields only D-(+)-glucose, indicating that these compounds are polysaccharides of D-glucose. Starch is comprised largely of the polymer with an α glycoside linkage (Figure 6.30). About 200 to 1,000 glucose molecules are linked together to form a starch molecule. The structure of cellulose is similar, except that the β glycoside linkage is involved and the molecular weight is higher, the molecule consisting of up to 14,000 glucose residues.

Fats such as lard and butter, and vegetable oils such as corn oil and olive oil, are composed principally of esters of the alcohol glycerol and various high-molecular-weight carboxylic acids (called fatty acids); esters of this type are usually referred to as *glycerides*:

$$
\begin{array}{l}
\text{H} \quad\quad\; \overset{\text{O}}{\overset{\|}{}} \\
\text{HC}-\text{O}-\text{C}-\text{R} \\
\quad\quad\;\; \overset{\text{O}}{\overset{\|}{}} \\
\text{HC}-\text{O}-\text{C}-\text{R}' \\
\quad\quad\;\; \overset{\text{O}}{\overset{\|}{}} \\
\text{HC}-\text{O}-\text{C}-\text{R}'' \\
\text{H}
\end{array}
$$

Table 6.2   *Fatty acids*

NAME OF THE ACID	FORMULA
Lauric	$CH_3(CH_2)_{10}CO_2H$
Myristic	$CH_3(CH_2)_{12}CO_2H$
Palmitic	$CH_3(CH_2)_{14}CO_2H$
Stearic	$CH_3(CH_2)_{16}CO_2H$
Oleic	$CH_3(CH_2)_7CH{=}CH(CH_2)_7CO_2H$ (*cis*)
Linoleic	$CH_3(CH_2)_4CH{=}CHCH_2CH{=}CH(CH_2)_7CO_2H$ (*cis,cis*)
Ricinoleic	$CH_3(CH_2)_5CH(OH)CH_2CH{=}CH(CH_2)_7CO_2H$ (*cis*)

The R groups in the glycerides may all be alike (R = R' = R''), or may be any combination; some of the more common fatty acids found in glycerides are given in Table 6.2. Most common fats are complex mixtures of glycerides containing several different fatty acids. In a few cases, however, a single fatty acid predominates and it can be recovered after hydrolysis. Myristic acid, for example, can be readily isolated from the fat of nutmeg, palmitic acid from coconut oil, and stearic acid from beef tallow.

Fats are the most calorie-rich of the foods, providing twice as many calories per gram as do the carbohydrates and the proteins. Fats have many functions in mammals, but one of the most interesting is their role in food storage for time of need; foodstuffs ingested in excess of the normal requirement are converted into fats and are deposited in certain areas of the body, to the chagrin of some human beings but to the advantage of animals that hibernate.

The biosynthesis of fats involves esterification of glycerol (as its phosphate) by the fatty acid (R'—CO_2H) in its activated form as a derivative of coenzyme A (R'—CO—SR).° The fatty acids themselves are formed by Claisen-type reactions similar to that outlined for the synthesis of citric acid, except that acetic acid (as acetyl CoA) is the sole carbon source. The long carbon chain of the fatty acid is thus built up two carbon atoms at a time (2 → 4 → 6 → 8 → . . .). This provides an explanation for the fact that the vast majority of fatty acids contain an even number of carbon atoms. Mixtures of normal alkanes have been isolated from fossils and from ancient sedimentary rocks. For example, the Green River shale (dating from the Eocene period; 50 million years old), has yielded a mixture of alkanes with odd numbers of carbon atoms (13, 15, 17, . . .); these are thought to have originated from the decarboxylation of the fatty acids, which as we have seen contain even numbers of carbon atoms.

The glycerides of the unsaturated fatty acids tend to have lower melting points than do the glycerides of saturated fatty acids; most oils are composed

° Check index for the structure of coenzyme A and for other references to this compound.

predominantly of the glycerides of unsaturated fatty acids. The hydrogenation of the double bonds of the fatty acids present in oils (converting them into saturated acid chains) raises the melting point of the glycerides. The "hardening" of vegetable oils such as corn and cottonseed oil to yield cooking fats such as Crisco and Spry is an important commercial process today:

$$\text{Oils} + H_2 \xrightarrow[\text{catalyst}]{\text{Ni}} \text{saturated fats}$$

An interesting application of this melting-point dependence on the degree of unsaturation is found in the fat composition of the sea anemone (*Metridium dianthus*). Sea anemones contain large amounts of fats, and in the species living off the shores of Florida, these fats are largely glycerides of saturated fatty acids. In contrast, related sea anemones living in colder waters off the coasts of New England contain fats made up largely of the unsaturated fatty acids; a Florida sea anemone deposited in the waters off New England would stiffen to the point of immobility.

LIPIDS   The term *lipids* refers to materials obtained from plant and animal sources that are soluble in oil but insoluble in water. The principal constituents of the lipids are fats, but a number of more complex glycerides are also represented. The phosphorus-containing lipids, or *phospholipids*, such as sphingomyelin, are important examples of complex pseudoglycerides; they are found principally in nerve and brain tissues.

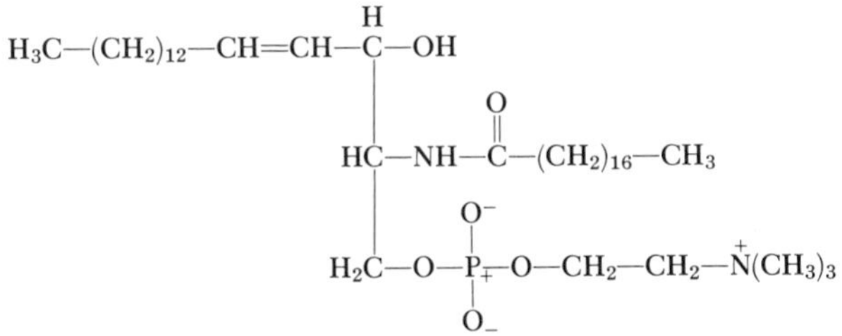

$$H_3C-(CH_2)_{12}-CH=CH-\overset{\overset{\displaystyle H}{|}}{C}-OH$$

Sphingomyelin

SOAPS   One of the most important of the chemical reactions of the glycerides is *saponification* (the hydrolysis with sodium hydroxide). The products are glycerol and sodium salts of the fatty acids present in the glyceride:

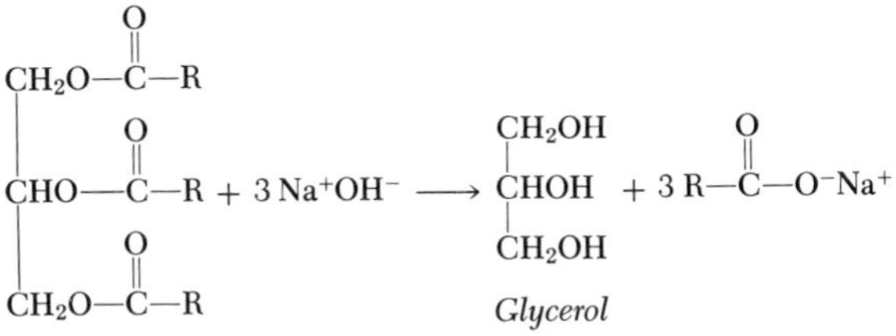

The sodium salts of long-chain fatty acids are called *soaps*. The soap of commerce, largely sodium stearate, is prepared by saponification applied to beef tallow. Before the twentieth century, which saw the ready availability of NaOH, soap was customarily prepared by the interaction of animal fats with water extracts of wood ashes (which contain sodium and potassium carbonates). This process (saponification brought about by the carbonates reacting as weak bases) has been in use since the days of the early Babylonians, and even today it is still used in primitive countries.

A soap molecule has two features that are essential for its cleansing action: a long hydrocarbon chain and a polar group (the carboxylate group). Modern detergents, such as sodium laurel sulfate, retain these two features.

$$CH_3(CH_2)_{10}CH_2-O-\overset{\displaystyle O}{\underset{\displaystyle O}{\overset{|}{\underset{|}{S}}}}-O^-Na^+$$

Sodium laurel sulfate

The wildflower *Saponaria officinalis* (also called bouncing Bet and soapwort) is an example of a group of plants used by the Indians and early pioneers that contain soaplike compounds. The polar group in these detergents is a group of sugar molecules bound by a glycoside linkage to the hydrocarbonlike group, which in these compounds is a steroid molecule.

OTHER NATURAL PRODUCTS

In addition to the three major classes of natural products, a number of other types exist, which, although they are not foodstuffs, are no less important in the proper functioning of living things. A full exposition of each type is beyond the scope of this volume; instead, a brief introduction to several types will be given, along with the formulas of representative members.

ALKALOIDS   Alkaloids are complex amines that are found in plants; the function of these compounds is unknown at the present time. Many of the alkaloids produce interesting physiological reactions in man (ranging from

Figure 6.31   **Alkaloids.**

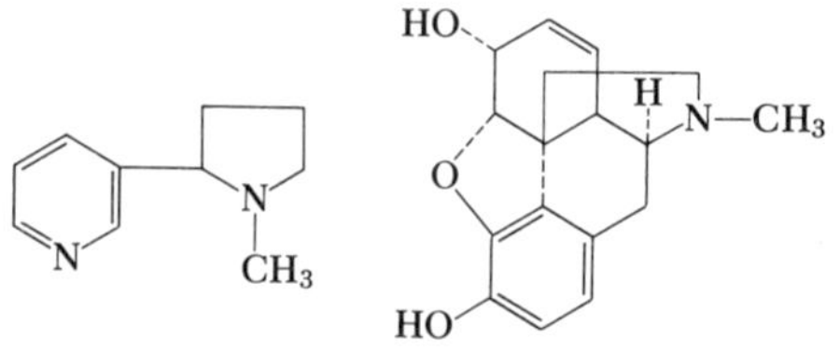

Nicotine
(tobacco leaves)

Morphine
(opium poppy)

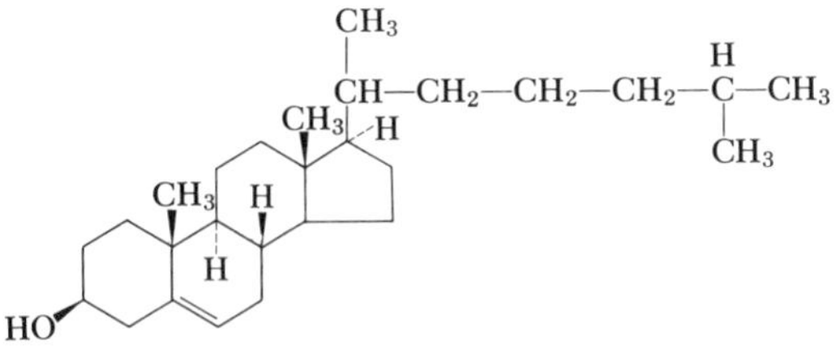

Figure 6.32 *Cholesterol,* $C_{27}H_{46}O$.

the relief of pain to the production of hallucinations) and for this reason, and also because alkaloids are easy to isolate, they were among the earliest natural products studied by chemists. Examples are given in Figure 6.31, along with the principal sources.

STEROIDS   Steroids are compounds containing a cyclopentanohydrophenanthrene ring system. The formula given in Figure 6.32 is that of cholesterol, the most common steroid in mammals; it has been estimated that a 140-lb man contains about 0.5 lb of cholesterol.

The steroids occur in most plants and animals, and in animals, at least, they are essential for the functioning of the organism. Members of this class include the sex hormones, the bile acids, the toad poisons, and also complex steroids that are used in treating arthritis and heart ailments. Compounds of this last group are often found attached to sugar molecules (as glycosides), and certain others contain alkaloids as integral parts of their molecules.

The steroid glycosides that affect heart action are toxic to higher animals when taken in other than very small doses. It is felt that this toxicity protects plants such as the common milkweed (*Asclepias syriaca*), which produce these compounds, from attack by herbivores. A certain group of butterflies has also taken advantage of the presence of these steroids. The monarch (*Danaus plexippus*), for example, lives exclusively on the plant juices of the milkweed and stores the toxic substances in its body. Most insectivorous birds have learned, from nonfatal trials, to avoid eating these butterflies, presumably using the shape and color of the butterflies as a cue. Still further along this chain of interrelationships are those nontoxic butterflies that have evolved so that their shape and color mimic precisely those of the toxic butterflies. Although the mimics do not feed on milkweed and related plants, they are nonetheless avoided by the birds.

TERPENES   The terpenes are compounds built up of 2-methylbutane units (C—C—C—C with a C branch). They have been isolated chiefly from plants, but certain important terpenes have also been isolated from animal sources. Squalene ($C_{30}H_{50}$), for example, is a terpene isolated from the liver of sharks; it is an intermediate in the synthesis of cholesterol in mammals. Two examples of

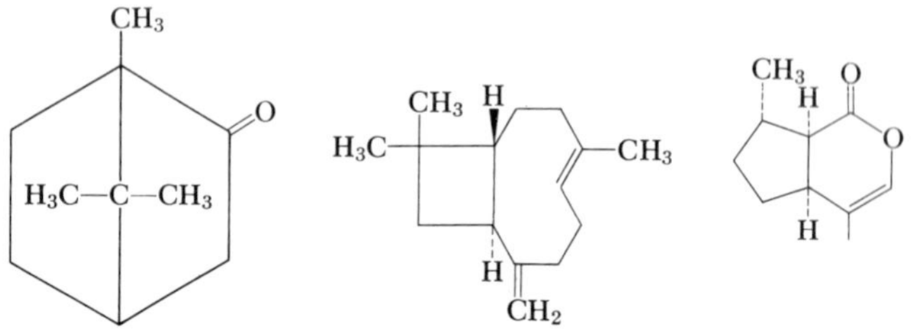

Camphor
(from the camphor tree,
Cinnamomum camphora)

Caryophyllene
(from oil of cloves)

Nepetalactone
(from catnip,
Nepeta cataria)

Figure 6.33   Terpenes.

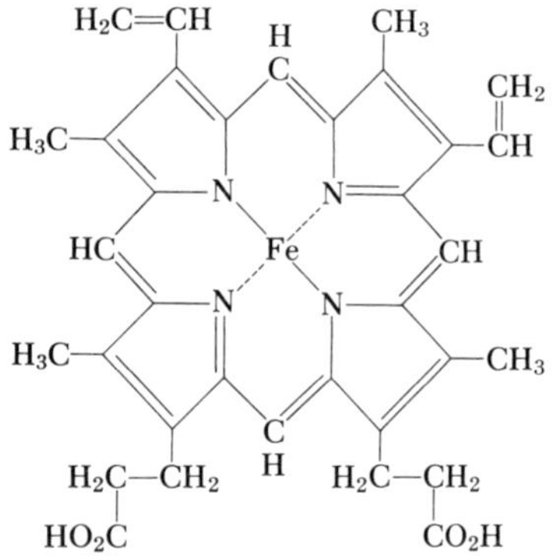

Figure 6.34   *The heme molecule.*

terpenes have been given elsewhere in this volume (Figure 3.13 and vitamin A, p. 94); other examples appear in Figure 6.33. Catnip has long been known to be highly attractive to members of the cat family, ranging from kittens to tigers. Until recently, there was no agreement as to why the active principle, nepetalactone, was produced in a plant. It has now been found that nepetalactone is a good insect repellant, and presumably the plant produces it for protection; the fact that felines like it is probably only incidental.

PORPHYRINS AND RELATED COMPOUNDS   Porphyrins are complex tetrapyrrole derivatives. The structure of heme, a member of this class, is given in Figure 6.34. Heme and a protein (globin) together make up the hemoglobin molecule, which is the oxygen carrier in blood. Chlorophyll, the key molecule in photosynthesis, is a related porphyrin containing magnesium (the structure of chlorophyll is given in the volume in this series on plant life[°]). Still other compounds related to heme, the cytochromes, are involved in electron transport and possibly in free-radical reactions in the cell.

VITAMINS   The vitamins are compounds essential for normal nutrition. Their structures vary widely, as is shown by examples cited elsewhere in the text (vitamin A, p. 94; pyridoxal phosphate, p. 123; and folic acid, p. 78). Only small amounts are required in the diet, a fact attributable to their biochemical role; most appear to be cofactors for enzymes, the cofactors being small molecules that bind to an enzyme and enable it to act as a catalyst. Catalysts (and cofactors) are not consumed by the reaction; small amounts are used over and over again by the organism in its normal metabolism.

ANTIBIOTICS   Antibiotics are commonly defined as compounds that inhibit the growth of bacteria. The structures of penicillin G, produced by a microorganism, and sulfanilamide, a synthetic antibiotic, are given in Figure 6.35. The latter compound is believed to be effective because of its structural

[°] A. W. Galston, *The Life of the Green Plant*, 2nd ed. (Englewood Cliffs, N.J.: Prentice-Hall, Inc., 1964).

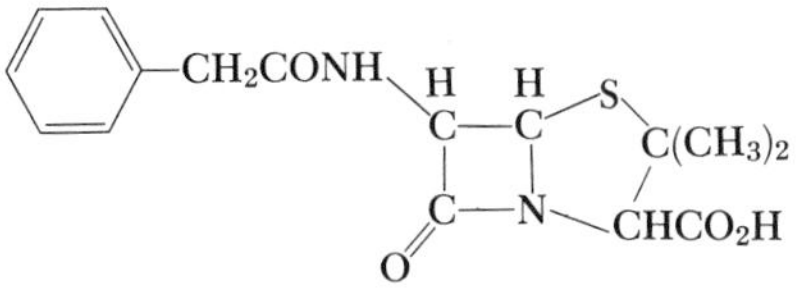

Penicillin G
(produced by the
mold Penicillium notatum)

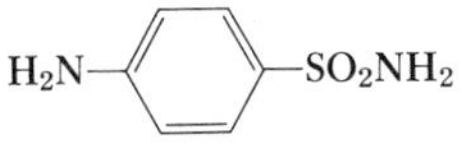

Sulfanilamide

Figure 6.35   **Antibiotics.**

resemblance to 4-aminobenzoic acid,

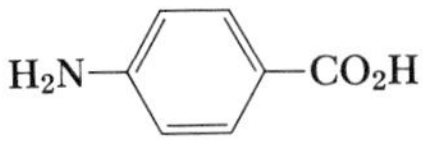

Because of this similarity, it competes with 4-aminobenzoic acid, which takes part in the formation of the vitamin folic acid (Figure 3.25); a sufficient amount will actually block the synthesis. In this way sulfanilamide can inhibit the growth of bacteria that assemble their own folic acid without having any marked effect on the human body since human beings must take in folic acid preformed in their food.

The chemistry of the natural products is usually subdivided further to include classes such as plant and animal hormones, nucleotides, compounds responsible for the coloring matter of flowers, and so on. It should be pointed out, however, that these are narrow divisions, and that many compounds are known that could fit equally well into several of these categories.

SELECTED READINGS

ATOMS, MOLECULES, CHEMICAL REACTIONS, AND INORGANIC CHEMISTRY

GRUNWALD, E., AND R. H. JOHNSON   *Atoms, Molecules and Chemical Change.* Englewood Ciffs, N.J.: Prentice-Hall, Inc., 1960. An introductory volume for those who have little or no background in science.

SIENKO, M. J., AND R. A. PLANE   *Chemistry,* 2nd ed. New York: McGraw-Hill Book Company, Inc., 1961. A good treatment of general chemistry at the college level.

ORGANIC CHEMISTRY

MORRISON, R. T., and R. N. BOYD   *Organic Chemistry,* 2nd ed. Boston: Allyn & Bacon, Inc., 1965. A good general introduction to organic chemistry.

FIESER, L. F., AND M. FIESER   *Advanced Organic Chemistry.* New York: Reinhold Publishing Corp., 1961. A thorough treatment of organic chemistry at the college level. Contains a number of chapters on natural products.

CRAM, D. J., AND G. S. HAMMOND   *Organic Chemistry,* 2nd ed. New York: McGraw-Hill Book Company, Inc., 1964. Organic chemistry organized in terms of reaction mechanisms.

WILSON, E. O.   "Pheromones," *Scientific American, 208,* Issue 5, p. 100 (1963). A popular account of the organic compounds used by insects in communication.

MCELROY, W. D.   *Cell Physiology and Biochemistry*, 2nd ed. Englewood Cliffs, N. J.: Prentice-Hall, Inc., 1964. A brief introduction to the subject of biochemistry.

BAKER, J. J. W., AND G. E. ALLEN   *Matter, Energy, and Life*. Palo Alto, Calif.: Addison-Wesley Publishing Company, Inc., 1965. A treatment of moderate length on biochemistry principles. Emphasis is placed on the physical nature of the bioreactions.

WHITE, A., HANDLER, P., AND E. L. SMITH   *Principles of Biochemistry*, 4th ed. New York: McGraw-Hill Book Company, Inc., 1968, A comprehensive treatment of biochemistry.

MOORE, W. J.   *Physical Chemistry*, 3rd ed. Englewood Cliffs, N.J.: Prentice-Hall, Inc., 1962. An interesting and comprehensive coverage of physical chemistry at the introductory level.

APPENDIX

A THE METRIC SYSTEM

LENGTH

1 meter (m) = 10 decimeters (dm) = 100 centimeters (cm) = 1,000 millimeters (mm)
1 cm = 10 mm
1 kilometer (km) = 1,000 m

MASS

1 kilogram (kg) = 1,000 grams (g) = 1,000,000 milligrams (mg)
1 g = 1,000 mg

VOLUME

1 liter = 1,000 milliliters (ml)
1 ml = 0.001 liter (l)

ENGLISH EQUIVALENTS

1 m = 39.37 inches (in)
2.54 cm = 1 in
1.61 km = 1 mile (mi)
1 kg = 2.20 pounds (lb)
28.35 g = 1 ounce (oz)
1 liter = 1.06 quart (qt)

Degrees Centigrade (°C) = 5/9(°F − 32)

Degrees Fahrenheit (°F) = 9/5 °C + 32

Degrees Absolute or Kelvin (°K) = °C + 273.2

GENERAL

The exponent of $10^2 = 2$, and 10^2 means 10 times $10 = 100$

$123,000 = 1.23 \times 10^5$

$1 = 10^0, 10 = 10^1, 100 = 10^2$, etc. $1/10 = 0.1, 1/100 = 0.01$, etc.

EXPONENTS

Multiplication (exponents are added):

$10 \times 10 \times 10 = 10^3 = 1,000$ $(8 \times 10^6)(9 \times 10^{-2}) = 72 \times 10^4 =$

$10^4 = 10 \times 10^3 = 100 \times 10^2,$ $7.2 \times 10 \times 10^4 = 7.2 \times 10^5$

   etc. $= 10,000$

Division (exponents are subtracted):

$1/10 = 10^0/10^1 = 10^{(0-1)} = 10^{-1},$ $\dfrac{7 \times 10^{-4}}{2 \times 10^3} = 3.5 \times 10^{-7}$

   $1/10^3 = 10^{-3}$, etc.   [that is, $(7 \times 10^{-4})(1/2 \times 10^{-3}) =$

                                         $7/2 \times 10^{-7}$]

Exponents of exponents (exponents are multiplied)

$(10^2)^3 = 10^6$ $\sqrt{10^4} = (10^4)^{1/2} = 10^{4/2} = 10^2$

Subtraction and addition (the exponents are made identical and the operation is performed):

$(4.5 \times 10^{-3}) - (2 \times 10^{-4}) = (45 \times 10^{-4}) - (2 \times 10^{-4}) = 43 \times 10^{-4} =$
4.3×10^{-3}

LOGARITHMS (TO BASE 10)

The logarithm of a number is the exponent to which 10 must be raised to equal the number:

$\log 1,000 = \log (10^3) = 3$ $\log 0.001 = \log (10^{-3}) = -3$

Multiplication (add logarithms):

$\log (100 \times 100) = \log (10^2 \times 10^2) = \log (10^2) + \log (10^2) = 2 + 2 = 4$
$\log 1,300 = \log (1.3 \times 10^3) = 0.11 + 3 = 3.11$ [the value 0.11 is obtained from a table of logarithms]

Division (subtract logarithms):

$\log (1,000/100) = \log (1,000) - \log (100) = 3 - 2 = 1$

INDEX